BrightRED Study Guide

CfE HIGHER

HUMAN BIOLOGY

Cara Matthew, Angela Grant and Kathleen Ritchie

BrightRED
PUBLISHING

First published in 2015 by:
Bright Red Publishing Ltd
1 Torphichen Street
Edinburgh
EH3 8HX

Reprinted with corrections in 2016

A CIP record for this book is available from the British Library.

ISBN 978-1-906736-64-4

With thanks to:
PDQ Digital Media Solutions Ltd, Bungay (layout), Anna Stevenson (copy-edit).

Cover design and series book design by Caleb Rutherford – e i d e t i c.

Acknowledgements
Every effort has been made to seek all copyright-holders. If any have been overlooked, then Bright Red Publishing will be delighted to make the necessary arrangements.

Permission has been sought from all relevant copyright holders and Bright Red Publishing is grateful for the use of the following:

Andrea Danti/Shutterstock.com (p 6); Mike Jones (CC BY-SA 2.5)[1] (p 7); ttsz/iStock.com (p 18); difught/iStock.com (p 33); OSTILL/iStock.com (p 33); koya79/iStock.com (p 39); Wolfgang Moroder (CC BY-SA 3.0)[2] (p 40); sportEX journals (CC BY-ND 2.0)[3]; Brocken Inaglory (CC BY-SA 3.0)[2] (p 60); chat9780/iStock.com (p 61); photo5963/iStock.com (p 61); selvanegra/iStock.com (p 73); ttsz/iStock.com (pp 75, 76 & 92); 4774344sean/iStock.com (p 77); susandaniels/iStock.com (p 77); blueringmedia/iStock.com (p 77); Suze777/iStock.com (p 79); dobrinya/iStock.com (p 80); vlad_karavaev/iStock.com (p 80); idrisesen/iStock.com (p 81); abadonian/iStock.com (p 81); leaf/iStock.com (p 82); choja/iStock.com (p 82); cako74/iStock.com (p 82); antikainen/iStock.com (p 83); MihailDechev/iStock.com (p 83); yangna/iStock.com (p 85); Minerva Studio/iStock.com (p 86); *Question taken from Higher Human Biology Exemplar, paper 2 question 12 © Scottish Qualifications Authority (p 89) (n.b. solutions do not emanate from the SQA).*

(CC BY-SA 2.5)[1]   https://creativecommons.org/licenses/by-sa/2.5/

(CC BY-SA 3.0)[2]   http://creativecommons.org/licenses/by-sa/3.0/

(CC BY-ND 2.0)[3]   http://creativecommons.org/licenses/by-nd/2.0/

Printed and bound in the UK by Charlesworth Press.

CONTENTS

INTRODUCTION

INTRODUCING CFE HIGHER HUMAN BIOLOGY

The CfE Higher Human Biology course is divided into four units:

- Unit 1: Human Cells
- Unit 2: Physiology and Health
- Unit 3: Neurobiology and Communication
- Unit 4: Immunology and Public Health

Units 1 and 2 are full units and Units 3 and 4 are half units.

ASSESSMENTS

To gain the CfE Higher Human Biology course award, you must pass all four units, as well as the Course Assessment.

Unit Assessments

- Each of the four units is assessed within your school using SQA Unit assessments.

- Practical abilities are also assessed internally. You are required to write a report of one of the investigations that you have carried out.

Course Assessment

The Course Assessment is made up of two components with a total of 120 marks. It will be graded A to D, which is determined on the basis of the total mark for both components.

Component 1: Question Paper

You will sit an externally assessed written examination consisting of a paper lasting 2 hours 30 minutes. It will be carried out under exam conditions and marked by SQA. This examination has an allocation of 100 marks and is divided into 2 sections:

1 Section 1 is the Objective Test which is worth 20 marks and consists of 20 multiple-choice questions.

2 Section 2 is worth 80 marks and contains restricted and extended-response questions. The extended-response questions each have a mark allocation of between 6 and 9 marks.

Marks for this written paper are distributed (approximately) proportionately across all four Units and the majority of the marks are allocated for applying knowledge and understanding. The remainder of the marks are awarded for applying scientific enquiry, analytical thinking and problem-solving skills.

Component 2: Course Assignment

The Course Assignment is worth 20 marks. You will investigate a relevant topic in biology that is related to one or more of the key areas in the Higher Human Biology course and then communicate your findings. This requires you to demonstrate your application of scientific-enquiry skills, and related biological knowledge and understanding.

Exam Hints

You do not need to answer the questions in order. At the beginning of the exam, find a question that you can answer easily, so that you settle your nerves.

Timekeeping is important, if you are to complete the whole paper. As a general rule, you should be taking just under one and a half minutes per mark. So, allowing ten minutes for settling at the start and checking your paper at the end, the timing for each section should be roughly:

contd

- Section 1, Objective Test: 25–30 minutes

- Section 2: approximately 1 hour 50 minutes.

Remember to look at the mark allocation for each question. Extended-response questions worth from 6 to 9 marks will require more lengthy answers – remember to allocate sufficient time to these.

Revision Tips

- Don't leave your revision until the last minute. Make up a revision schedule, giving yourself enough time to revise thoroughly, and stick to it. Be realistic – you should work around your other activities and remember that you do need to take time off to relax.

- Find somewhere to study that is quiet and uncluttered. You need space to spread out your work.

- Study for short periods (between 30 and 45 minutes) with short breaks in between, to keep up your level of concentration. Go out of the room where you are studying during each break; you will return refreshed and ready for your next study session.

- Read over each sub-topic at a slower pace than you would usually do, asking yourself questions or reading aloud. Make sure that you understand what you have been reading – you only learn what you understand.

- It's often easier to remember facts if you talk about topics with a family member or a friend. So, find a study buddy who can ask you questions about your work.

- Practice makes perfect; do past-paper practice so that the exam format is as familiar as possible. There are only a few ways in which you can be asked the same question, and you will see similar questions and diagrams appearing in many past papers. Doing a past paper against the clock will also help you to get your time management right.

- In the run up to the exams, eat plenty of fresh fruit and vegetables to keep your energy levels up, and make sure that you get a good night's sleep so that you are alert throughout the exam.

- Switch off all mobile devices and social media.

THE STRUCTURE AND AIMS OF THIS BOOK

There is no shortcut to passing any course at Higher level. To obtain a good pass requires consistent, regular revision over the duration of the course. The aim of this revision book is to help you achieve success by providing you with a concise and engaging coverage of the CfE Higher Human Biology course material. We recommend that you use this book in conjunction with your class notes, to revise each topic area, prepare for Unit Assessments, and other internal assessments, and in your preparation for the final exam.

The book is divided between the four units of the course. Within each section, there is a double-page spread on each of the sub-sections.

Each double-page spread:

- provides the key ideas and concepts of the sub-section in a logical and digestible manner

- contains 'Internet Links' and 'Don't Forget' boxes that flag up vital pieces of knowledge that you need to remember and important things that you must be able to do

- gives a link to an online test to test your knowledge and understanding of each topic

- ends with a 'Things To Do and Think About' feature which extends your knowledge and understanding of the subject, and provides additional interest. Sometimes there are questions to help you check your understanding.

Good luck, and enjoy!

HUMAN CELLS

DIVISION AND DIFFERENTIATION IN HUMAN CELLS

DIFFERENTIATION

As soon as fertilisation occurs, the **zygote** (fertilised egg) begins to divide by mitosis. It forms a ball of identical and **unspecialised** cells. Then, as the cells become specialised for specific functions, regions in the continually dividing embryo start to behave differently from each other. This is called differentiation and happens when:

- **some genes** are switched off

- genes that are vital to all living cells, for example those for respiratory enzymes, are **expressed** (transcribed)

- genes for specialised cell functions are expressed. For example, contractile proteins are synthesised in muscle, the enzymes required to make neurotransmitters are synthesised in neurons, and the digestive enzymes are synthesised in specific regions of the digestive system.

The somatic (body) cells increase in number and renew by mitosis.

STEM CELLS: AN OVERVIEW

Stem cells are undifferentiated and can divide to form specialised cells. They are found in all parts of the body and can differentiate into one or a few cell types that are characteristic for their location. The degree of specialisation that occurs depends on when and where these cells are found.

Early Embryo Stem Cells

The inner cell mass of the early **embryo** is **pluripotent**, which means the cells that are found there are capable of dividing to form *all types* of specialised cell.

Organs are made from a combination of tissues which are groups of specialised cells. For example, the heart is made of muscular, nervous, epithelial and connective tissues.

Dissected heart

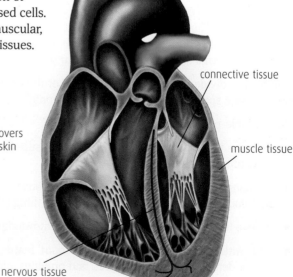

connective tissue

muscle tissue

nervous tissue

An early embryo (blastocyst)

nervous tissue forming nerves

epithelial tissue which covers the outer surface forming skin and inner body cavities and vessels.

muscular tissue made of muscle cells

inner cell mass

connective tissue forming blood, bone and cartilage

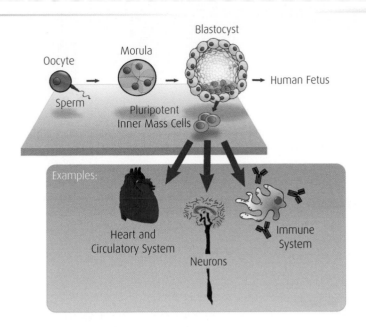

Embryonic Stem Cells

Embryonic cells tend to divide into specialised cells; they do not naturally **self-renew** (replace) themselves. They can, however, be treated in a lab under *in vitro* conditions to form stem cells, which are known as **embryonic stem cells**. This provides a bank of stem cells for research.

Adult Stem Cells

These are found in specific locations, **replacing** specialised cells that have a limited lifespan or that are damaged, or producing more cells to enable the **growth** of the tissue. Adult stem cells are multipotent, which means that their ability to differentiate is **limited** to the tissue they come from.

Some adult stem cells do not become specialised but **self-renew** (divide to make more stem cells) in order to maintain a stock to last the organism's lifetime.

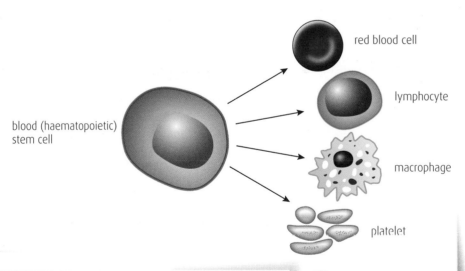

blood (haematopoietic) stem cell

red blood cell

lymphocyte

macrophage

platelet

 THINGS TO DO AND THINK ABOUT

1. What types of genes are expressed in a specialised somatic cell?

2. Explain the difference between pluripotent and multipotent stem cells?

3. Why do some adult stem cells self-renew?

STEM CELLS

GERMLINE CELLS

Germline cells are situated in the testes and ovaries. They can replace themselves by dividing by **mitosis** or they can produce haploid gametes by dividing by **meiosis**.

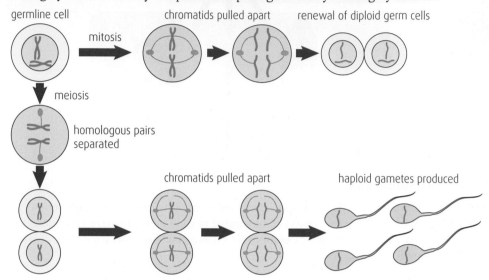

If a mutation occurs in a germline cell it will end up in gametes and may be passed on to offspring if fertilisation is successful.

STEM CELL RESEARCH

Stem cell research gives a better understanding of the control of gene expression and differentiation. Stem cells are also used to model cells and tissues, to study the effects of diseases or drug therapy. A limitation of this research is that it can't look at interactions within the organism. A key aim is to promote the therapeutic use of stem cells to replace damaged or diseased tissue. Stem-cell production is under strict control.

RESEARCH AND THERAPEUTIC USES OF STEM CELLS

Therapeutic value looks at the potential of stem cells in medicine. Stem cells are of special interest in the repair of diseased or damaged organs and to replace lost tissue.

Skin Grafts

If a person is badly burned, there may not be enough good skin to use in grafts. A solution is to remove some adult stem cells from an area of good skin. These can then be cultured in the laboratory to produce skin cells. The new skin is grafted onto affected areas on the patient. The skin will not be rejected but isn't perfect as it lacks the complexity of normal skin, not having hair follicles and sweat glands.

Bone Marrow Transplantation

Bone marrow stem cells are multipotent and can produce blood cells and platelets. Transplantation is used to treat certain types of blood-related cancers, leukaemia or sickle cell anaemia. The patient's own bone marrow stem cells are destroyed and then replaced with healthy bone marrow stem cells, either from a compatible donor or from the patient themselves (if healthy bone marrow was harvested previously).

contd

Cornea Repair

Damaged corneas are more usually removed and replaced with healthy corneas from dead donors. This procedure is tried and tested, but is invasive and there is a global lack of donors. Scientists have found that multipotent stem cells are located at the edge of the cornea. They can produce corneal or conjunctival cells. Stem cells can be removed from the patient's 'good' eye to be transplanted onto the damaged eye. Studies have included the use of contact lenses as culture media for the stem cells.

ETHICAL ISSUES OF STEM CELL USE

Certain control measures are in place to try to address some of the ethical implications of stem cell use.

Moral

Unused blastocysts from embryonic stem cell lines are destroyed as they are not allowed to develop beyond day 14. This is when the embryo would normally implant in the uterus leading to development of a foetus.

Health

A complete medical history of adult stem cell donors is required to minimise the chance of recipients developing other medical problems.

Safety

Stem cells must be safe to use in the treatment of patients: they should not cause other conditions or diseases such as tumours. It is for this reason that ongoing research and thorough testing is vital.

CANCER

Normal cells have a programmed lifespan and are replaced by cell division when they die, so that organs maintain both shape and function. A cancer cell does not undergo this pre-programmed 'death' but divides rapidly to form a space-occupying mass or tumour.

Healthy cells spend most of their lifecycle growing and performing their specialised roles. A short portion of their lifecycle is spent in mitosis. Checkpoints in the lifecycle ensure that that the cell has grown sufficiently and completed DNA replication before division. A cell undergoes a limited number of cell divisions before it dies. Some genes produce regulatory proteins that promote division and act like an accelerator, others stop or slow division and act like the brakes.

Cancer cells lack these controls because of mutations in the genes that control mitosis. The cells divide rapidly and form a tumour. A cancerous tumour has a blood supply that feeds the cells with nutrients and oxygen for rapid growth. Healthy cells normally stick to each other, but cancer cells lose this ability and separate. The blood vessels that supply them allow cells to escape to form secondary tumours.

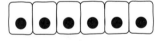
normal tissue

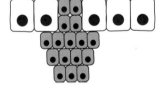
tumour

 VIDEO LINK

Watch the animation on cancer formation at www.brightredbooks.net

 ONLINE TEST

Test yourself on your knowledge of stem cells at www.brightredbooks.net

 THINGS TO DO AND THINK ABOUT

1   Where are germline cells located?
2   What is produced when germline cells divide by:
    a   mitosis?   b   meiosis?
3   Describe the mechanisms that cause the formation and spread of cancerous tumours.

DNA

Deoxyribonucleic acid (DNA) contains the code to make all of the tens of thousands of proteins in an organism. Proteins, in the form of enzymes, catalyse the reactions to manufacture a complete individual that is correct for its species. A gene is a region of DNA that codes for the specific sequence of amino acids that forms a polypeptide chain.

The polypeptide chain can be modified and folded to form a protein. This genetic code is passed on through generations when gametes fuse during fertilisation. Thus, the genetic code is inherited.

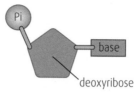

deoxyribose

sugar-phosphate backbone

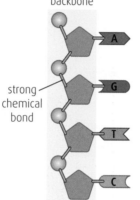

strong chemical bond

THE STRUCTURE OF DNA

DNA is made up of sub-units called nucleotides, joined in strands. There are four types of nucleotide, depending on the base: adenine (A), thymine (T), cytosine (C) and guanine (G). Each strand is made up of nucleotides which form strong chemical bonds between a phosphate group of one nucleotide and the deoxyribose of another nucleotide: the sugar–phosphate backbone. The DNA molecule is double-stranded due to the formation of weak hydrogen bonds between the bases: adenine always bonds with thymine (A–T) and cytosine always bonds with guanine (C–G).

The strands are anti-parallel, meaning that they run in opposite directions. Each strand has a 3' end and a 5' end, determined by whether the third or fifth carbon on the sugar molecule of the nucleotide is closest to the end. The double-stranded DNA molecule twists to form a double helix.

CHROMOSOMES

The DNA is located in the membrane-bound nucleus and takes the form of chromosomes. The DNA in each chromosome is extremely long and thread-like; it must, therefore, be organised into tidy spools (a bit like spools of thread), so it cannot get tangled up. Each spool is composed of proteins with the DNA tightly coiled around them. This is how 2 metres of DNA is packed into the microscopic nucleus of every cell in the human body!

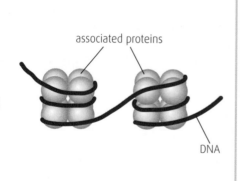

associated proteins

DNA

5'   hydrogen   3'
bonds

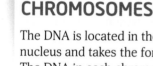

3'   5'

VIDEO LINK

The video clip at www. brightredbooks.net gives a nice animation of our tightly coiled and packed DNA.

DON'T FORGET

Nuclear DNA is tightly coiled with associated proteins to form linear thread-like chromosomes.

DNA REPLICATION

DNA must be replicated (duplicated) before cell division can occur, ensuring daughter (new) cells have a complete set of genetic information. Mitosis is the type of division that replaces diploid body cells; meiosis is the type that produces haploid gametes.

STAGES OF DNA REPLICATION

Requirements for replication:

- DNA
- ATP
- DNA polymerase (enzyme)

- the four types of nucleotide
- primer (a short sequence of nucleotides)
- ligase (enzyme)

contd

1 The DNA molecule **unwinds**.
2 Hydrogen bonds break, '**unzipping**' the molecule and exposing the bases of both DNA strands.
3 A **primer** attaches to one end of each exposed DNA template strand.
4 This initiates **DNA polymerase** to add free complementary DNA nucleotides to the **3'** end of the growing strand.
5 **Hydrogen bonds** form between the bases.
6 Strong **chemical bonds** form between the phosphate and deoxyribose sugar of adjacent nucleotides.
7 **Ligase** enzyme joins the fragments to form a complete strand.
8 Each replicated DNA molecule is made of one original template strand and a newly synthesised strand.

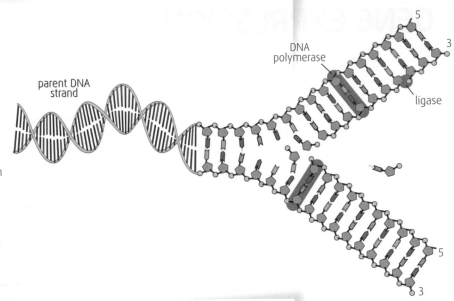

DIRECTION OF REPLICATION

DNA polymerase adds nucleotides on to the 3' end of the strand. The strands run in opposite directions, which means that one strand can be built up continuously as the molecule unzips; exposing a 3'-end nucleotide – this is the **leading strand**.

However, the other strand unzips the wrong way to add nucleotides to the 3' end. Synthesis using this strand lags behind until enough of the template strand is exposed to add a primer and then add nucleotides to the 3' end. This is repeated as more template becomes exposed, producing fragments. A complete strand is made when the fragments are joined together by the enzyme **ligase**. This strand is the **lagging strand**.

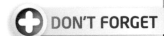

VIDEO LINK

Find out more about mitochondrial DNA at www.brightredbooks.net

 THINGS TO DO AND THINK ABOUT

1 A molecule of DNA was found to be composed of 32% adenine. Express the ratio of thymine to guanine as a simple whole-number ratio.

2 Label the following on the diagram of a section of DNA:

   a The molecules represented by numbers 1–4.

   b Show the 3' and 5' ends by adding labels to the blank boxes.

   c Name the type of bonds labelled A and B.

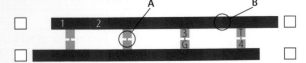

3 Complete this diagram to show the direction of DNA replication. Show the 3' and 5' ends and draw the leading strand with a continuous line and the lagging strand with a broken line.

DON'T FORGET

Each strand can only be replicated in the 5' to 3' direction.

ONLINE TEST

Take the test on DNA online at www.brightredbooks.net

GENE EXPRESSION

PHENOTYPE

Gene expression determines the phenotype of an organism. The **proteins** produced (enzymes, hormones, structural and transport proteins) all work together to determine the characteristics that are typical for a species. A specialised cell, such as a skeletal muscle cell, expresses two types of gene: those that are vital for its maintenance, such as the genes coding for respiratory enzymes, and those that are vital for its specialised function, for example the genes coding for the slow- and fast-twitch muscle fibres. All other genes are switched off. Characteristics that are unique to the individual (such as height and body mass) are the result of genotype and are influenced by intracellular and extracellular environmental factors. Diet, activity levels, stress levels and infection can all affect an organism's internal environment, triggering changes in pH, chemical and hormone production. This, in turn, affects gene expression by switching genes on or off.

RNA

Ribonucleic acid (RNA) is a single-stranded molecule made of nucleotide sub-units. Each nucleotide consists of a phosphate group, ribose sugar and a base. There are four different bases: adenine, uracil, guanine and cytosine.

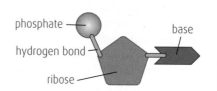

Strong chemical bonds form between the phosphate group of one nucleotide and the ribose of the next nucleotide to form an RNA molecule.

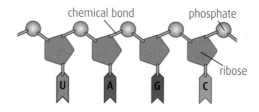

You should be familiar with the following types of RNA, all of which are involved in protein synthesis.

- **Messenger RNA (mRNA)** is formed during transcription of DNA in the nucleus and is the template for protein synthesis at the ribosomes. It dictates the sequence of amino acids in the protein being synthesised.

	DNA	mRNA
Type of sugar	Deoxyribose	Ribose
Bases	Adenine, cytosine, guanine and **thymine**	Adenine, cytosine guanine and **uracil**
Number of strands	Two	One
Location	Only in nucleus	Moves from nucleus to cytoplasm

Comparison of DNA and mRNA

contd

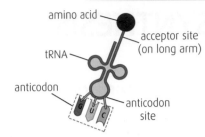

amino acid

acceptor site
(on long arm)

tRNA

anticodon

anticodon
site

Transfer RNA (tRNA) carries specific amino acids to the ribosomes for translation of the genetic code. Base pairing within tRNA molecules causes the tRNA to fold into a clover-leaf shape with two distinct regions: an exposed triplet anticodon site and an attachment site for a specific amino acid. The anticodon matches to its complementary triplet codon on the mRNA strand, bringing its specific amino acid with it.

Ribosomal RNA (rRNA) binds to proteins to form ribosomes. A ribosome is composed of two sub-units, one large and one small and is the site of protein synthesis; the small sub-unit 'reads' the code on mRNA and the large sub-unit provides active sites so that two tRNAs can bring their amino acids next to each other allowing peptide bonds to form between them.

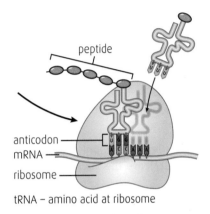

peptide

anticodon
mRNA

ribosome

tRNA – amino acid at ribosome

DON'T FORGET

RNA is transcribed from DNA. mRNA is translated into protein but tRNA and rRNA are not translated into protein.

ONLINE

Find out more about RNA at www.brightredbooks.net

ONLINE TEST

Revise your knowledge of this topic by taking the test at www.brightredbooks.net

 THINGS TO DO AND THINK ABOUT

1 Compare the structures of DNA and mRNA.

2 Summarise the roles of the three types of RNA.

3 Draw and label a molecule of tRNA.

PROTEIN SYNTHESIS

TRANSCRIPTION

The first stage of protein synthesis takes place in the nucleus and is called transcription. An mRNA molecule is produced that carries the genetic code from the DNA in the nucleus to a ribosome in the cytoplasm. Production of mRNA is essential, as DNA is too large to pass through the nuclear membrane.

The process is as follows:

1   DNA unwinds as RNA polymerase moves along a section that codes for a protein.

2   The DNA molecule 'unzips' when hydrogen bonds are broken.

3   Bases on the DNA strands are exposed.

4   mRNA nucleotides move in and form complementary base pairs with one of the DNA strands (the coding strand). Weak hydrogen bonds form. Cytosine always pairs with guanine; adenine on DNA pairs with uracil on mRNA, and thymine on DNA pairs with adenine on mRNA.

5   Strong chemical bonds form between the phosphate of one nucleotide and the ribose of the next nucleotide, building the mRNA strand.

6   The weak hydrogen bonds that were holding the DNA and mRNA strands together break, allowing the mRNA primary transcript to leave the nucleus and enter the cytoplasm.

7   Hydrogen bonds reform between the two DNA strands, and the DNA molecule rewinds to form a double helix.

DON'T FORGET

A triplet of bases on mRNA is called a codon, and codes for one amino acid.

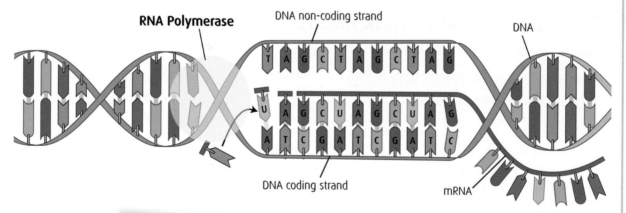

RNA SPLICING

The original DNA template contains the code to make protein. However, it also contains non-coding sections.

The functions of different non-coding regions of DNA are:

- to regulate transcription – there are binding sites for other chemicals that can switch off neighbouring genes

- to be transcribed to form RNAs that are not translated into protein, for example tRNA, rRNA and RNA fragments

- unknown – some regions make no sense and are thought to be the result of mistakes.

DON'T FORGET

The functions of non-coding regions are not mandatory knowledge in your CfE Higher Human Biology course but it is useful to understand why they are present.

contd

The coding regions are called **exons** and the non-coding regions are called **introns**.

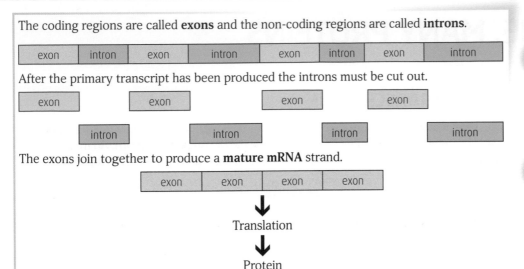

exon	intron	exon	intron	exon	intron	exon	intron

After the primary transcript has been produced the introns must be cut out.

| exon | exon | exon | exon |

| intron | intron | intron | intron |

The exons join together to produce a **mature mRNA** strand.

| exon | exon | exon | exon |

↓

Translation

↓

Protein

✚ DON'T FORGET

Translation is initiated by a start codon and ends with a stop codon.

◐ VIDEO LINK

Watch the clip at www.brightredbooks.net for an overview of protein synthesis that includes the roles of different RNA molecules.

TRANSLATION

The mRNA molecule that is formed during transcription leaves the nucleus via a nuclear pore and attaches to a **ribosome**. The ribosome is made up of two units composed of ribosomal RNA (rRNA) and proteins. The nucleic code is now ready to be translated into an amino acid sequence.

1   Translation is initiated by the start codon (AUG) on the mRNA strand.
2   Transfer RNA (tRNA) molecules become attached to amino acid molecules in the cytoplasm. Each type of amino acid attaches to a specific tRNA molecule.
3   tRNA molecules in the cytoplasm transport amino acids to the ribosome.
4   The first tRNA molecule moves in by means of base pairing between the anticodon on the tRNA molecule and the complementary codon on the mRNA strand.
5   Another tRNA molecule carries an amino acid to the ribosome. Complementary pairing between codon and anticodon brings the amino acids in line beside each other. A peptide bond forms between the amino acids.
6   The first tRNA molecule detaches from the mRNA and is free to collect another amino acid from the cytoplasm.
7   As translation progresses, the ribosome moves along the mRNA molecule exposing the third codon, allowing a third tRNA molecule to bring a third amino acid into position.
8   This process repeats until the stop codon (UGA) is reached at the 3' end of the mRNA strand and the newly formed polypeptide chain is complete.

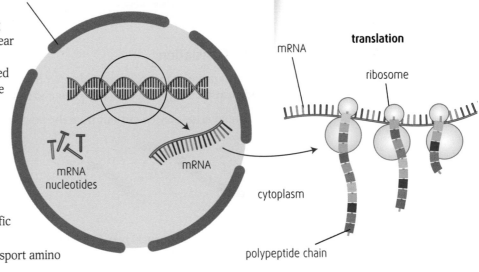

💭 THINGS TO DO AND THINK ABOUT

1   Write the complementary mRNA code to this strand of DNA:
    AGGCTAACTGCAATCGAAATG
2   List all the raw materials required for transcription to take place.
3   Describe the sequence of events leading to transcription.
4   What is the difference between an intron and an exon?
5   How does a primary transcript differ from a mature mRNA strand?

✓ ONLINE TEST

Head to www.brightredbooks.net and test yourself on this topic.

ONE GENE, MANY PROTEINS

AFTER TRANSCRIPTION

One **gene** may code for more than one protein. The codes for different proteins may be made by cutting out alternative regions (splicing) of the **mRNA transcript**. An **intron** (non-coding region) for one protein may be an **exon** (coding region) for a different protein.

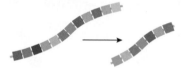

mRNA transcript

Splicing option 1	Splicing option 2	
exon intron **exon** intron exon exon	intron exon **intron** exon intron intron	gene
intron **intron**	intron **intron** intron intron	non-coding regions
exon exon exon exon	exon exon	splicing regions
exon **exon** exon exon	exon **exon**	spliced mRNA
↓	↓	
Leaves nucleus to be translated into Protein 1	Leaves nucleus to be translated into Protein 2	

After Translation

Different proteins may also be made *after* translation during **post-translational modification**. Polypeptides may be adapted in several ways after transcription.

1   Amino acids may be **cut out** from the polypeptide chain.

2   The polypeptide chain may be **combined** with another polypeptide chain. (For example, insulin is made of two polypeptides.)

human insulin

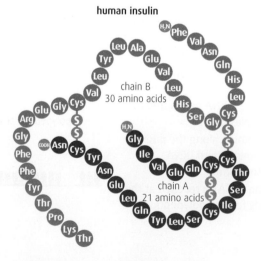

3   A **phosphate** group might be added.

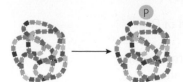

contd

4 A **carbohydrate group** might be added.

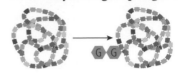

STRUCTURE OF PROTEINS

Once the primary structure (sequence of amino acids) has formed, the polypeptide becomes coiled and is held together by hydrogen bonds.

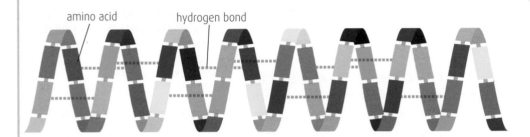

amino acid   hydrogen bond

Finally, the polypeptide forms sheets (in fibrous proteins) or is wound up to form a ball (in globular proteins). The three-dimensional shape is determined by **hydrogen bonds** that hold the chains together. The protein's shape is due to interactions between amino acids, which cause it to twist and fold in specific ways; hydrophobic areas of the chain move inwards and away from the watery external environment, and positively and negatively charged amino acids move towards each other.

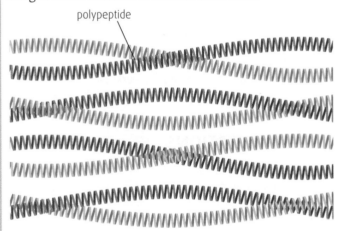

polypeptide

fibrous proteins – flat sheets

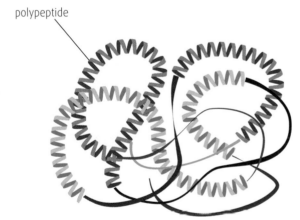

polypeptide

globular proteins – wound into a ball

THINGS TO DO AND THINK ABOUT

1 What type of bond is responsible for:
   a connecting amino acids together in the primary structure
   b coiling of the polypeptide chain
   c forming the three-dimensional shape of the protein.

2 How does post-transcriptional modification result in different proteins?

3 Describe how post-translational modification may result in different proteins.

MUTATIONS

A **mutation** can cause a change in the amino acid sequence, or in the number of amino acids in a protein. The resultant protein may be either **absent** (because it no longer functions) or **faulty**, causing **genetic disorders**.

SINGLE-BASE MUTATIONS

Individual genes are affected by **point mutations**. If the sequence of DNA bases in a gene is altered, the corresponding sequence of amino acids may change, possibly altering the protein produced. Think about the mutations in the diagram.

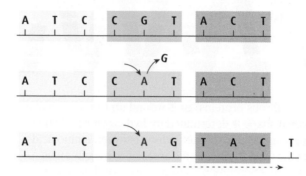

| A | T | C | C | G | T | A | C | T |

Normal strand of DNA

| A | T | C | C | A | T | A | C | T |

Substitution – one of the bases is replaced by a different base (here A replaces G). This can result in **splice site, nonsense** and **missense** mutations.

| A | T | C | C | A | G | T | A | C | T |

Insertion – an extra base is inserted into the sequence, in this example A, moving the bases after the insertion one place to the right. This is a **frame-shift mutation**, as every amino acid after the point of insertion is altered.

| A | T | C | C | T | A | C | T | A |

Deletion – a base is removed from the sequence, in this example G, shifting the bases one place to the left. This is another **frame-shift mutation**, as every amino acid after the point of deletion is altered.

CONSEQUENCES OF POINT MUTATIONS

DNA triplets code for specific amino acids. There are 20 amino acids but 64 triplets (from the possible combinations of the four nucleotide bases A, T, C and G arranged in sets of three). So, an amino acid may be coded by several triplets. This means that a mutation can have any one of five consequences:

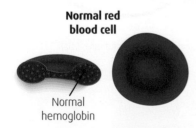

Normal red blood cell

Normal hemoglobin

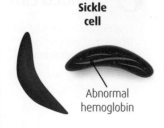

Sickle cell

Abnormal hemoglobin

1 **Silent:** a nucleotide is substituted, but the new triplet codes for the same amino acid and the protein is normal.
2 **Neutral:** the substituted nucleotide results in a triplet that codes for a different but similar amino acid, and the mutated protein functions normally.
3 **Missense:** the substituted nucleotide results in a triplet that codes for a different amino acid, which changes the function of the protein. This is seen in sickle-cell disease – red blood cells are malformed (sickle-shaped), reducing their oxygen-carrying capacity.
4 **Nonsense:** the substituted nucleotide results in a stop codon, so the polypeptide chain is shorter. This occurs in Duchenne muscular dystrophy, which causes a progressive weakening of muscles and reduced life expectancy, mainly in males.
5 **Frame-shift:** this is the result of a deletion or insertion mutation and the entire sequence of triplets after this point will be wrong. An example of a condition caused by this type of mutation is Tay-Sachs syndrome; a progressive disease of the nervous system which becomes obvious at 6 months of age. Affected children have a life expectancy of 3 to 5 years.

SPLICE SITE MUTATIONS

A mutation may alter codons that trigger the splicing of introns, so these are not cut out but remain in the mRNA sequence. These introns are translated and the protein has extra amino acids, altering its function. Alternatively, mutations may give rise to codons for exon–intron splicing, causing the mRNA to be cut in the wrong place. Beta-thalassemia affects the structure of haemoglobin and is caused by this type of mutation. People with this condition have recurring anaemia and require blood transfusions.

NUCLEOTIDE-SEQUENCE REPEAT EXPANSION

A substitution mutation could mean that this sequence:

CAGCCGCAG is changed to: CAGCAGCAG giving a repeat of the CAG sequence.

CAG repeats are diagnostic of Huntington's Disease. This is a degenerative neurological disorder that shows up in middle-age and is both life changing and life shortening. The greater the number of repeats of CAG, the earlier the onset of the condition.

VIDEO LINK

Learn more about mutation by watching the clips at www.brightredbooks.net

DON'T FORGET

Mutations affect proteins: by altering the amino acid sequence and thereby the protein function; by causing the loss of a protein; by duplicating or altering regulatory instructions, giving excess protein, or by creating different combinations of proteins.

CHROMOSOME STRUCTURE MUTATIONS

Whole sections of chromosomes can be altered by mutation, affecting several genes.

Duplication	Deletion	Translocation
Genes from one of a pair of homologous chromosomes transfer to the other, leading to duplication of genes. Myoglobin and haemoglobin are thought to have evolved from a common ancestral gene that duplicated and, subsequently, mutated.	Genes are lost from a chromosome.	Sections are swapped between different chromosomes.

GENETIC DISORDERS CAUSED BY CHROMOSOME MUTATIONS

Mutations involving whole segments of chromosomes are often lethal, but some cause conditions that have an impact on the individual's life.

Mutation	Genetic disorder	Consequences to health
Deletion on chromosome 5	Cri-du-chat syndrome	Some degree of learning disability, distinctive facial features, normal life expectancy.
Translocation of genes between chromosome 22 and chromosome 9	Chronic myeloid leukaemia (CML)	Slow progressing leukaemia. Individuals have to take medication for life and may need a bone marrow transplant.

THINGS TO DO AND THINK ABOUT

1 Redraw these chromosomes to illustrate the following chromosome structure mutations:
   a translocation
   b duplication
   c deletion.

2 A section of DNA has the following sequence: ATGCAGTAC. What type of gene mutation is represented by each of the following altered sequences?
   a ACGCAGTAC
   b ATGCATAC
   c ATGCATGTAC

ONLINE TEST

How well have you learned about mutations? Test yourself at www.brightredbooks.net

HUMAN GENOMICS

All of the genes in an organism are known collectively as the **genome**, and the study of the genome is known as **genomics**. Technology has enabled scientists to determine the nucleotide sequence of specific genes and the entire genomes of certain species.

A comparison of sequence data has revealed that different species are surprisingly similar, which suggests that the genome is **highly conserved**.

SEQUENCING DNA

DNA can be sequenced using automated computer analyses – a branch of science known as **bioinformatics**.

Each chromosome contains many million nucleotide base pairs but sequencers can only determine a few hundred at a time. The solution is to make copies of the DNA to be sequenced.

Copy 1 is broken up into small fragments. Copy 2 is also broken into fragments, but in different places:

1 DNA Copy 1 is cut into fragments with enzyme X (note it cuts in one place)

TACCGTCGAATTCCGTAAGCTA   TTTTCAGTAACTAAGTAACTTCATCC

2 DNA Copy 2 is cut into fragments with enzyme Y (note it cuts in two places).

TACCGTCG   AATTCCGTAAGCTATTTTCAGT   AACTAAGTAACTTCATCC

3 Each fragment is sequenced.

Copy 1
TTTTCAGTAACTAAGTAACTTCATCC
TACCGTCGAATTCCGTAAGCTA

Copy 2
AACTAAGTAACTTCATCC
TACCGTCG
AATTCCGTAAGCTATTTTCAGT

4 A computer analyser builds up the complete sequence by identifying the start fragment , which begins with TAC, then matching the Copy 1 and Copy 2 fragments by looking for overlapping sequences.

The codon TAC marks the start codon, so we know this is the beginning. The yellow copy TAC fragment is longer than the TAC blue fragment. So, we know the second blue fragment must start with the overlapping sequence AATTCCGTAAGCTA.

TACCGTCG|AATTCCGTAAGCTA
Overlap-----------
TACCGTCG|AATTCCGTAAGCTA|TTTTCAGT

The end of this blue fragment also contains the sequence TTTTCAGT which corresponds to the start of the second yellow fragment. In this way, the sequence is deduced.

TACCGTCGAATTCCGTAAGCTA TTTTCAGT AACTAAGTAACTTCATCC
TACCGTCGAATTCCGTAAGCTATTTTCAGT AACTAAGTAACTTCATCC
Start   overlap area | overlap area

The sequence, therefore, is:

TACCGTCGAATTCCGTAAGCTATTTTCAGTAACTAAGTAACTTCATCC

DNA sequencing generates massive amounts of data; the smallest chromosome in the human complement contains over 51 million nucleotide base pairs and there are about three billion nucleotides in the human genome. The data is stored in specialist **databases** around the world, such as the National Center for Biotechnology Information, NCBI, or the Human Genome Mutation Database, HGMD. The databases are available via the **internet** and are accessed by scientists for further **analyses** and **comparison**.

ONLINE

Explore the human genome by following the NCBI link at www.brightredbooks.net

USES OF BIOINFORMATICS

Finding Gene Sequences

This can be achieved by:

- matching a sequence to a similar sequence in a database
- finding a **start** codon and sequencing until the **stop** codon is identified
- working backwards to identify all the possible DNA sequences from the amino acid sequence in a protein.

Other species of importance to humans have been targeted for DNA sequencing: crop plants, farm animals, crop pests and organisms that cause disease in humans or food species. The information may be used, for example, to develop genetically enhanced crops or to find alternatives to the chemical control of agricultural pests and pathogens.

ONLINE

Use the link at www.brightredbooks.net to study the protein linked to Retinitis Pigmentosa, a condition affecting vision. You will follow the protein's story from gene sequence to structure.

PHYLOGENETICS

Phylogenetics compares sequence data of different species to determine how related they are and when they diverged (split) from a common ancestor. In phylogenetic diagrams, the length of the branch is proportional to the number of mutations in the sequence.

Species C has more differences in sequence, so is most distantly related to the common ancestor.

Species A and B have more similarities with each other, so diverged from each other later.

DON'T FORGET

Bioinformatics is the use of automated computer analyses to determine the sequence of DNA. The infomation has to be stored in vast databases.

PERSONAL GENOMICS

The human genome sequence was completed in 2003 and provides a reference database. Individuals all have slight variations in their sequences, which can provide information about their risk of developing certain diseases or conditions, the potential effectiveness of certain medicines or the risk of developing serious side effects of treatment. Assessing risk is a complex task as mutations are not always harmful.

An individual's personal genome sequence could, therefore, be used to predict the risk of developing conditions, allowing them to make lifestyle choices to reduce the risks; it could also be used to inform the prescription of the most effective drugs. This type of personalised medicine is known as pharmacogenetics.

It is important to note that the development of a disease is often due to a combination of genetics and modifiable factors, such as diet, activity and stress levels. Genetic information could, therefore, be misinterpreted if taken out of context. There are also ethical implications regarding the right of access to genetic information by prospective employers and insurers.

 THINGS TO DO AND THINK ABOUT

Use the link at www.brightredbooks.net to identify the amino acid sequence of this section of DNA code:

ACGTGCTGCCGACGAGGCATCCGAAACCTTCTTTAG

How many amino acids does the sequence encode?

What does the final triplet stand for?

Press the **RELATE FUNCTION** button to find out which proteins contain this sequence.

ONLINE TEST

Take the test on genomic factors at www.brightredbooks.net

POLYMERASE CHAIN REACTION (PCR)

DNA replication occurs naturally in cells before cell division. Given the correct conditions, DNA fragments can be amplified, through repeated cycles of artificial replication, in laboratories by a process known as the polymerase chain reaction.

STAGES OF PCR

Requirements:

- an original DNA sample to provide the template
- a stock of four DNA nucleotides
- heat-tolerant DNA polymerase (an enzyme)
- a thermal cycler (an automated reaction vessel)
- buffer solution (maintains an optimum pH)
- copies of a DNA primer for the start of each strand of the target fragment (which targets the fragment to be amplified).

Heat-tolerant (thermostable) polymerase is obtained from bacteria that live around hot springs and which are able to withstand high temperatures without their proteins being denatured.

PCR PROCESS

1 DNA is heated to 95°C to separate the original DNA strands.

2 The sample is cooled to 55°C. Complementary primers are added and these bind (anneal) to the start of each strand.

3 The sample is heated to 72°C and heat-tolerant DNA polymerase is added.

4 Complementary free DNA nucleotides are added to the 3' end of the new strands.

5 The number of original molecules has now doubled – this is called amplification.

Steps 1–5 are repeated, amplifying the DNA to make many copies.

DON'T FORGET

A primer is needed to start the synthesis of a DNA strand. Primers are short nucleotides sequences that are complementary to the start of the DNA sequence to be copied. They are needed because DNA polymerase can only build a strand by adding nucleotides to the end of an existing sequence.

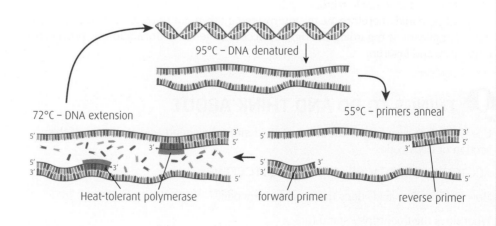

95°C – DNA denatured

72°C – DNA extension

55°C – primers anneal

Heat-tolerant polymerase   forward primer   reverse primer

GENE PROBES

Gene probes are used to find target sequences in DNA.

Gene probes are short, synthetic, single strands of DNA which are complementary to the sequence of interest. These probes need to be identifiable, so have other compounds attached, such as fluorescent labels.

Fluorescent Labelling

Such fluorescent probes can be incubated with whole cells or chromosomes, and these viewed with a fluorescence microscope. Often the fluorescence is weak; the solution is to amplify DNA fragments using PCR to give a larger sample and, therefore, a stronger fluorescence. The fragments are separated by electrophoresis: samples are loaded into wells in a gel and an electric current applied, which separates the fragments according to size and charge. Any fragments that contain the target sequence will be fluorescent and can be seen under a UV light.

probe   fluorescent label

ACCGACTAG

TCGTAATTGGCTGATC DNA fragment

PRACTICAL APPLICATIONS OF PCR

1 DNA Profiling

If only a tiny quantity of genetic material is found at a crime scene, PCR can provide enough material for various tests, such as genetic fingerprinting. Each individual has unique repeats of sequences. Specific probes are incubated with the crime-scene sample and with a sample from a suspect. If comparison of the two samples provides a match, there is a high degree of certainty that the suspect has been identified. This same method can be used to identify unknown victims or dead persons by comparing their DNA with close relatives.

2 Research

PCR can be used to build up a large bank of material from a small initial source, such as a mutated gene sequence. The stock can then be subjected to several types of analyses and can even be divided up so that several research teams can work on it.

3 Early Detection of Infection

PCR can be used to amplify viral DNA when only a few cells out of millions are infected, such as in the case of early HIV infection.

4 Diagnostics

Amplified stocks of DNA can be tested for a known genetic condition. If a mutation is found, the individual can be given advice on their chances of developing the disease.

Other diagnostic methods include:

1 **Archaeobiology**: the DNA found in mummified bodies, skeletal remains or insects trapped in amber is usually highly degraded. However, small intact samples may be amplified and sequenced to look for connections to other mummified remains, find living relations or build a picture of life at that time.

2 **Phylogenetics**: this maps when species split from each other in evolution by how much their DNA differs. Again, small samples from mummified or fossil evidence can be used.

THINGS TO DO AND THINK ABOUT

1 Explain why it is necessary to increase the temperature to 92°C during a PCR cycle, then cool it to 55°C.
2 Heat-tolerant DNA polymerase is obtained from bacteria that live in the margins of hot springs and are adapted to withstand a temperature range of 50 to 80°C. What would happen if you used DNA polymerase from bacteria that live in a temperature range from −5 to 45°C?

VIDEO LINK

Learn more about PCR by watching the clip at www.brightredbooks.net

DON'T FORGET

PCR gives researchers large stocks of target DNA that can be tested in a variety of ways or stored for future research.

ONLINE TEST

Test your knowledge of PCR at www.brightredbooks.net

METABOLISM

METABOLIC PATHWAYS

Within the cell, reactions can be classed as:

- **anabolic** reactions – large molecules are synthesised from several smaller ones, with energy being used up
- **catabolic** reactions – large molecules are broken down into smaller molecules, usually with a release of energy.

The sum of all the anabolic and catabolic reactions that occur within a living cell is collectively known as the cell's **metabolism**.

A metabolic pathway is a series of chemical reactions that follow on, one after another. Each stage in the pathway is controlled by an enzyme, with the product of one reaction becoming the substrate for the next. The reactions are often **reversible**. In the example shown below, compound X will be converted into compound Y, as long as the concentration of X remains relatively high compared to that of compound Y. If the concentration of X decreases, the reaction will proceed in the opposite direction. In this way, enzymes act to drive reactions towards equilibrium, where the relative concentrations of molecules at the start and end of the reaction are balanced.

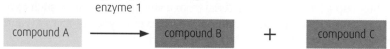

Some reactions are **irreversible**, which means the reaction will not go in the opposite direction. This is usually because the reaction has gone to completion and the products are stable, meaning they will no longer react with each other.

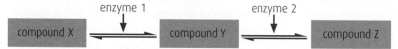

Chemicals can take part in more than one reaction. This means that they can take **alternative routes**, as in the diagram below.

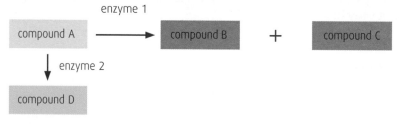

CONTROL OF METABOLIC PATHWAYS

Where several enzymes are involved in catalysing different stages of a metabolic pathway, the enzymes may be situated close together in a cell membrane, producing a **multi-enzyme complex**. By locating close together within the membrane, the overall rate of reaction of the whole pathway can be increased, making the pathway more efficient. DNA polymerase and RNA polymerase both form part of multi-enzyme complexes.

SWITCHING GENES ON AND OFF

As all cells inherit a complete set of genetic information during mitosis, each cell has the potential to produce protein from every gene in the code. However, different types of cells only make the proteins required by that cell type for normal function. This means that there must be a mechanism which allows genes to be 'switched on' and 'switched off'.

Signal Molecules

Signal molecules can trigger reactions that switch genes on or off. Some are secreted by the cell and work elsewhere in the body – these are called **extracellular signal molecules**. Hormones, which travel in the blood circulation to reach their target cells, are examples.

Intracellular signal molecules work inside the cell that produced them. Two scientists, Jacob and Monod, suggested a hypothesis (the Jacob–Monod hypothesis) to explain this, after studying the bacterium *Escherichia coli* (*E. coli*). The bacterium synthesises an enzyme called β-galactosidase, which catalyses the breakdown of the milk sugar lactose into glucose and galactose. Jacob and Monod found that β-galactosidase is only produced when a bacterium detects that lactose is present, preventing unnecessary use of resources and energy by the cell. So, how does a bacterium control expression of the gene for β-galactosidase?

The diagram below shows the genes that are involved.

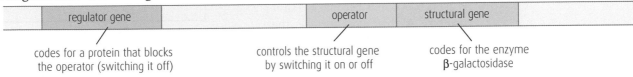

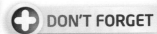

Lactose Absent

When lactose is absent, a repressor molecule (coded for by the regulator gene) binds with the operator. Because the operator gene is switched off, the structural gene remains switched off, preventing transcription of the genetic code for the enzyme β-galactosidase.

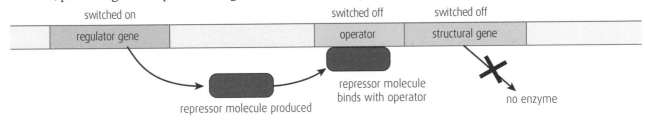

Lactose Present

When lactose is present, the repressor molecule binds to lactose, instead of to the operator. The operator now switches on the structural gene, allowing transcription of the code for β-galactosidase, and the enzyme is manufactured.

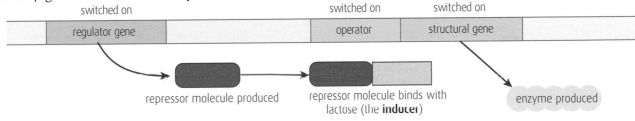

When all the lactose has been broken down, the repressor molecule binds to the operator gene, switching it off again.

REGULATION OF THE RATE OF REACTION OF ENZYMES

The rate of reaction can be regulated by the availability of substrate or the concentration of product.

 THINGS TO DO AND THINK ABOUT

1 Describe the functions of the following gene types:
   a regulator   b operator   c structural.
2 Name two advantages of having regulator genes.

ENZYME ACTION

ACTIVATION ENERGY

At normal body temperature and without enzymes, chemical reactions would take place at too slow a rate to maintain life. Chemical reactions involve breaking chemical bonds and forming new ones. To start a reaction, energy (the activation energy) must be used to break bonds within the reactant molecules. As energy is absorbed, bonds become increasingly unstable. At their most unstable, the molecules are said to be in a **transition state**. When the bonds in the reactants break, the molecular structure of the product forms. Enzymes act by reducing the activation energy required to reach the transition state and, therefore, they allow reactions to take place at a lower temperature.

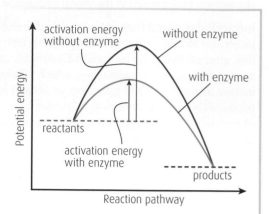

INDUCED FIT

Each enzyme can only act on one substrate. Enzyme action is, therefore, said to be **specific**. This is because the shape of the substrate molecule fits into the enzyme's active site. When two or more substrate molecules are involved in the reaction, the molecules fit into the active site in the particular orientation required to allow the molecules to react.

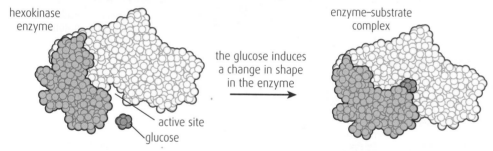

Enzyme function is described as an 'induced fit'. The active site begins in an open position, which allows the substrate (with a high affinity for the active site) to move in and bind. Binding of the substrate causes a change in the shape of the active site to a closed position. This brings the substrate and enzyme closer together, increasing the chance of a reaction. Once the product has formed, the shape of the active site returns to the open position and the product (which has a low affinity for the active site) moves out.

DON'T FORGET

To increase the rate of reaction further, the enzyme concentration would have to be increased.

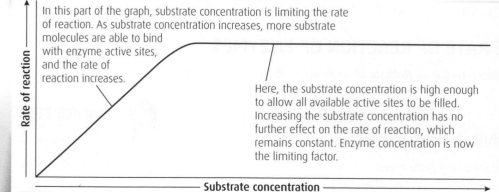

In this part of the graph, substrate concentration is limiting the rate of reaction. As substrate concentration increases, more substrate molecules are able to bind with enzyme active sites, and the rate of reaction increases.

Here, the substrate concentration is high enough to allow all available active sites to be filled. Increasing the substrate concentration has no further effect on the rate of reaction, which remains constant. Enzyme concentration is now the limiting factor.

SUBSTRATE CONCENTRATION

As the substrate concentration increases, the rate of reaction increases and then becomes constant. The graph demonstrates how the rate of reaction is linked to the number of active sites that are filled at any one time.

FEEDBACK INHIBITION (END-PRODUCT INHIBITION)

The rate at which some metabolic pathways progress can be controlled by a build-up of the end product. In feedback inhibition, the end product binds to one enzyme in the metabolic pathway, altering the shape of this enzyme's active site and stopping the pathway. This prevents too much end product from being produced. As the concentration of the end product drops, inhibition ceases and the pathway resumes again.

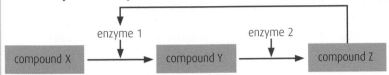

Some molecules of compound Z bind to enzyme 1, inhibiting its function.

COMPETITIVE INHIBITION

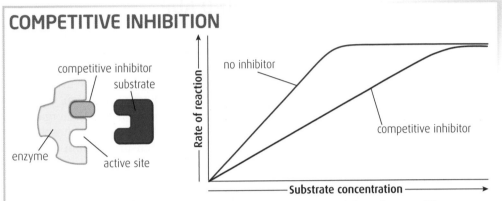

Competitive inhibitor molecules have a shape similar to that of the substrate. They bind with the enzyme's active site, preventing the substrate from entering. Because the substrate and inhibitor are in competition for the active site, increasing the substrate concentration causes an increase in the rate of reaction.

NON-COMPETITIVE INHIBITION

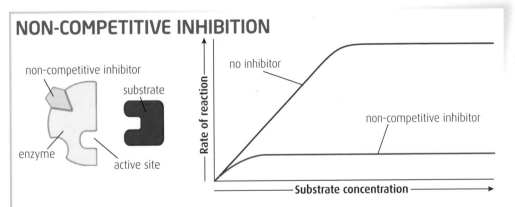

Non-competitive inhibitors bind to a part of the enzyme which is not the active site. As a result, the shape of the active site is altered and the substrate cannot enter. Because the substrate and inhibitor are not in competition for the active site, increasing substrate concentration has no effect on the rate of reaction. The rate of reaction remains low.

 THINGS TO DO AND THINK ABOUT

1 Describe the induced fit theory of enzyme action.

2 Describe the mechanisms of competitive and non-competitive inhibition.

3 Explain why increasing substrate concentration cannot overcome the effect of non-competitive inhibition.

 DON'T FORGET

Cyanide is a non-competitive inhibitor that inhibits aerobic respiration.

 VIDEO LINK

Check out the clip at www.brightredbooks.net for more on enzymes.

 ONLINE TEST

Take the test on enzymes at www.brightredbooks.net

CELLULAR RESPIRATION 1

RESPIRATORY SUBSTRATES AND USES OF ENERGY

Molecules which can be broken down to release energy in respiration are called respiratory substrates. The energy that is released is used to fuel cellular processes, such as protein synthesis, contraction of muscle, active transport and DNA replication.

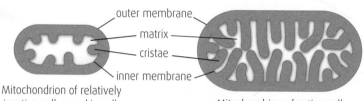

outer membrane
matrix
cristae
inner membrane

Mitochondrion of relatively inactive cell, e.g. skin cell

Mitochondrion of active cell, e.g. muscle cell

ATP AND ADP

The series of reactions that make up respiration result in chemical energy being transferred to a molecule called ATP, adenosine triphosphate. ATP is a source of energy that can be used immediately by cells. During respiration, ATP is made when a bond forms between an inorganic phosphate (Pi) and ADP, adenosine diphosphate. This reaction is called **phosphorylation**. When the bond is subsequently broken, the energy is released and used in cellular processes, such as synthetic pathways.

Mitochondria

Mitochondria are known as the power houses of the cell because they are the main site of ATP synthesis.

STAGE ONE – GLYCOLYSIS

Glycolysis takes place in the cytoplasm of every living cell. No oxygen is required. It is divided into an **energy investment phase** and an **energy pay-off phase**.

Initially, the conversion of $2ATP \rightarrow 2ADP + 2Pi$ provides energy for the conversion of glucose to an intermediate phosphorylated molecule (phosphate groups are added to the glucose molecule). The first phosphorylated molecule can take part in alternative reactions. The second phosphorylated molecule is the product of an **irreversible** reaction catalysed by the enzyme **phosphofructokinase** and is committed to the glycolysis reaction. Phosphofructokinase is, therefore, important in directing glycolysis.

This phase is followed by the energy pay-off phase in which the intermediate molecules are converted into two pyruvate molecules. As four ATP molecules are produced during this phase, there is a **net gain of two ATP** (4ATP produced, 2ATP used) in glycolysis. Hydrogen ions are released during the energy pay-off phase and are picked up by the hydrogen carrier molecule NAD to make NADH.

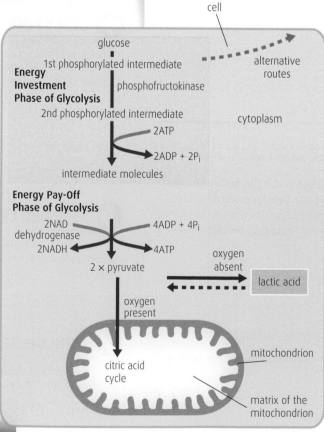

LACTIC ACID METABOLISM

In the absence of oxygen, lactic acid metabolism takes place. Lactic acid metabolism produces only the two ATP molecules released in glycolysis. Pyruvate is converted to lactic acid.

ONLINE TEST

How well have you learned about cellular respiration? Take the test online at www.brightredbooks.net

DEHYDROGENASE

Dehydrogenase enzyme removes **hydrogen** and **electrons** from respiratory intermediates and passes them to the hydrogen carriers NAD to form NADH and FAD to form $FADH_2$. These molecules are involved in further energy generation.

STAGE TWO – CITRIC ACID CYCLE

The citric acid cycle takes place in the matrix of mitochondria. Breakdown of pyruvate produces carbon dioxide and an acetyl group. The acetyl group binds with co-enzyme A to produce acetyl co-enzyme A. The acetyl group and oxaloacetate then combine to produce citrate. As the cycle proceeds, carbon dioxide is released and hydrogen ions are picked up by either NAD or FAD to form NADH and $FADH_2$ respectively. NADH and $FADH_2$ carry the hydrogen ions to the third stage of respiration, the electron transport chain, on the inner membrane of the mitochondrion.

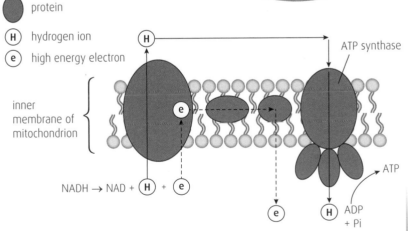

ELECTRON TRANSPORT CHAIN

The electron transport chain is found on the inner membrane of the mitochondria and consists of a series of carrier proteins. Hydrogen ions and electrons are transferred from **NADH** and **FADH$_2$** to the electron transport chain. As the electrons are passed down the chain, energy is released. Electrons are finally passed to oxygen, which binds with hydrogen ions in the matrix to produce water. Energy that is released from the electron transport chain is used to pump hydrogen ions (protons) from the matrix to the intermembrane space, causing a proton concentration gradient to develop. Hydrogen ions return to the matrix by flowing through a channel in the enzyme ATP synthase. This flow of ions causes parts of the enzyme molecule to rotate in a clockwise direction, changing the shape of the active site and allowing the conversion of ADP + Pi to ATP. A total of 34 ATP molecules is made in the electron transport chain from each glucose molecule respired.

THINGS TO DO AND THINK ABOUT

1 Describe the role of phosphofructokinase in directing the respiratory pathway.

2 What is the role of dehydrogenase?

DON'T FORGET

One molecule of glucose yields 2ATP from glycosis; 2ATP from the citric acid cycle and 34 ATP from the electron transport chain giving a total yield of 38 ATP.

CELLULAR RESPIRATION 2

ENERGY SOURCES IN RESPIRATION

Glucose is the main respiratory substrate. However, other carbohydrates, fats and proteins can be broken down and used as respiratory substrates when required.

CARBOHYDRATES	glucose → used in glycolysis
	starch and glycogen → broken down into glucose for use in respiration
	sugars other than glucose → converted into either glucose or intermediate molecules for use in glycolysis
FATS	Fat contains twice the energy of either carbohydrate or protein. It can be broken down into fatty acids and glycerol, both of which can enter the respiratory pathway.
	fatty acids → converted to acetyl co-enzyme A, before entering the citric acid cycle
	glycerol → converted to intermediate molecules for use in glycolysis
PROTEINS	While excess dietary protein can be used as an energy source, most of the protein that is taken in is used for growth and repair of body tissues.
	Proteins are broken down to produce amino acids. When excess amino acids are broken down, some of the products enter the respiration pathway.
	proteins → amino acids → converted to pyruvic acid, acetyl co-enzyme A, or intermediates in the citric acid cycle

During exercise, muscle cells use the energy source that is most readily available to them. In sequence, this means: muscle glycogen, plasma glucose, muscle fatty acid, then plasma fatty acids. The concentration of fatty acids in the blood plasma is increased by the breakdown of fat stores in the body. Carbohydrate is the preferred substrate during high-intensity, short-burst exercise, such as sprinting, while fat metabolism is increased during lower intensity but endurance exercise, such as jogging or marathon running.

Proteins are a resource of last resort and their use involves the breakdown of muscle as there is no other store of this substrate. This is seen in starvation and causes a loss of muscle mass. The heart is a muscular sac, so there may be serious consequences to the starving individual's long-term health.

REGULATION OF RESPIRATION

All chemical pathways must be regulated to ensure that valuable resources are not wasted in making unnecessary products.

ATP is the final product of respiration and, when it is plentiful, it prevents more from being produced by inhibiting the enzyme **phosphofructokinase**. This regulatory mechanism is known as **feedback inhibition**.

Citrate also **inhibits** the action of phosphofructokinase when it is plentiful. When ATP is in demand, the citrate molecules are being used up, leaving phosphofructokinase uninhibited and free to drive glycolysis. In this way, glycolysis and the citric acid cycle slow down when there is little need for ATP, as there is an accumulation of citrate and ATP. If citrate is running out, because of a sudden high demand for ATP, it can no longer inhibit phosphofructokinase. This means that glycolysis speeds up, supplying more **acetyl groups** to the citric acid cycle, thus generating more citrate and ultimately more ATP.

The concentration of citrate determines the rate of respiration and **synchronises** it to match the demand for ATP.

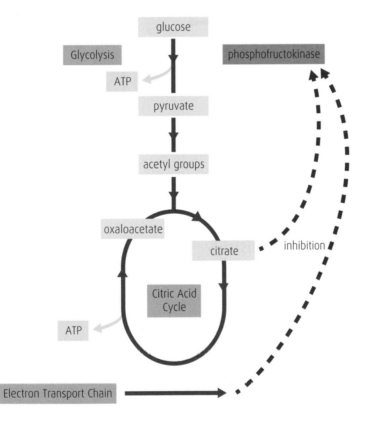

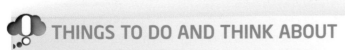

THINGS TO DO AND THINK ABOUT

Describe the role of citrate and ATP in regulating respiration.

EXERCISE

STRENUOUS EXERCISE

ATP is required for muscle contraction. Strenuous exercise means that ATP stocks in muscle cells become depleted, as the demand for oxygen outstrips supply. There is an alternative means of regenerating ATP, but this is only good for a short **10-second burst** of energy and it utilises the muscle cell compound called **creatine phosphate**. This can be broken down to release phosphate and energy to convert ADP to ATP.

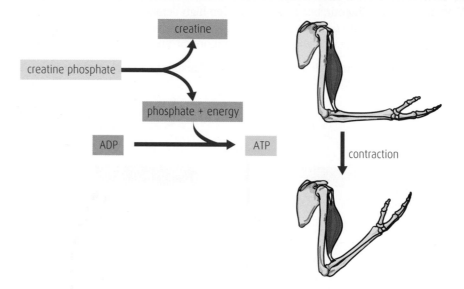

Creatine phosphate reserves are replenished when exercise stops, normal respiration resumes and the need for energy is low.

SKELETAL MUSCLE FIBRES

Skeletal muscles move the joints of the skeleton. They are made up of a mixture of slow- and fast-twitch muscle fibres.

Slow-Twitch Muscle Fibres

These are useful for endurance sports, such as cycling and long-distance running. They rely on aerobic respiration to supply their ATP and, as these organelles are the site of aerobic respiration, contain many mitochondria. Additionally, they have a rich blood supply to deliver the oxygen required to maintain an aerobic environment. The fibres contain large quantities of myoglobin. Myoglobin is similar to haemoglobin and is a protein that stores and transports oxygen. The prefix 'myo' means muscle and this protein is only found in muscle. The main source of energy in slow-twitch fibres is fat.

Fast-Twitch Muscle Fibres

These muscle fibres are needed for sudden bursts of energy, which is required in such sports as the javelin, weight lifting and shot putt, for example. These cells rely on creatine phosphate and glycolysis to supply their ATP requirements and, as a result, possess fewer mitochondria. They also have fewer blood vessels as they rely on glycolysis and do not need such an efficient supply of oxygen. Their main sources of energy is glycogen and creatine phosphate.

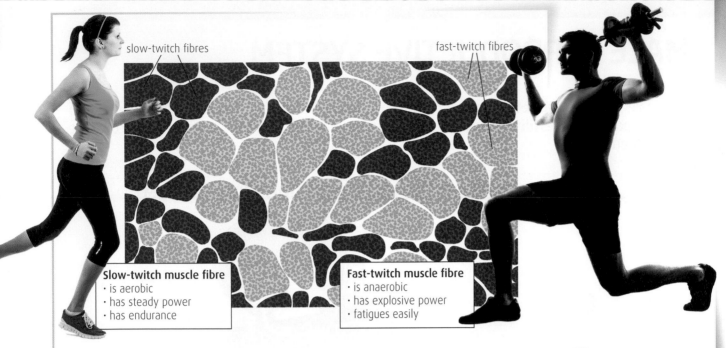

slow-twitch fibres

fast-twitch fibres

Slow-twitch muscle fibre
• is aerobic
• has steady power
• has endurance

Fast-twitch muscle fibre
• is anaerobic
• has explosive power
• fatigues easily

VIDEO LINK

See a video on slow-twitch and fast-twitch muscle fibres at www.brightredbooks.net

THINGS TO DO AND THINK ABOUT

1 Why are slow-twitch muscle fibres darker than fast-twitch muscle fibres?

2 Predict the ratio of slow-twitch to fast-twitch fibres in a middle-distance runner.

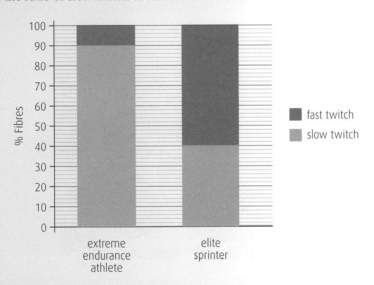

■ fast twitch
■ slow twitch

DON'T FORGET

Fast-twitch muscle fibres have fewer mitochondria and slow-twitch muscle fibres have many mitochondria.

3 Summarise the difference between slow- and fast-twitch muscle fibres.

	Slow-twitch fibres	Fast-twitch fibres
Number of mitochondria		
Blood supply		
Source of energy		
Type of activity		

PHYSIOLOGY AND HEALTH

MALE REPRODUCTIVE SYSTEM

The male produces **gametes** (**sperm cells**) continuously from puberty until death and is said to display **continuous fertility**. This is possible because male sex hormone levels remain constant after puberty.

VIDEO LINK

Learn more about sperm production by watching the clip at www.brightredbooks. net

ORGANS OF THE MALE REPRODUCTIVE SYSTEM

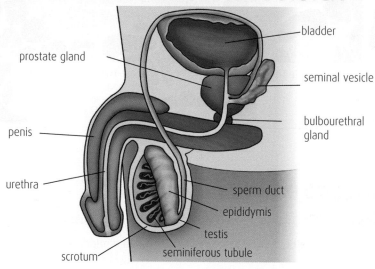

Sperm Production

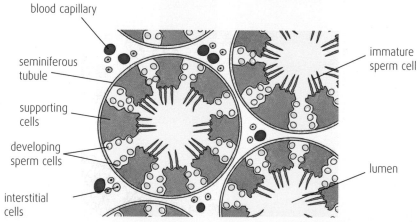

Sperm cells are produced in the **seminiferous tubules** inside the **testes**. These coiled tubes are surrounded by blood vessels and clusters of **interstitial cells** that produce the hormone **testosterone**. Meiosis of **germline cells** in the wall of the seminiferous tubules gives rise to immature sperm cells, which are released by **supporting cells** into the lumen of the seminiferous tubule.

Sperm Cells

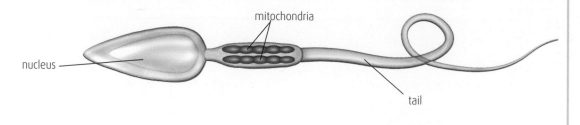

SEMEN

During ejaculation, **semen** passes through the sperm duct and the urethra to be ejected into the female reproductive tract. Semen is a mixture of sperm and secretions from the **seminal vesicles** and **prostate gland**, which maintain the mobility and viability of the sperm. The reproductive glands contribute to semen as follows:

Prostate gland	The prostate gland secretes a milky fluid containing **enzymes** that keep the fluid thin.
Seminal vesicles	The seminal vesicles secrete an alkaline, viscous fluid containing **fructose** and **prostaglandins**. Fructose is a respiratory substrate which provides energy for movement of the sperm tail. Prostaglandins cause the female reproductive tract to contract, helping sperm movement.

DON'T FORGET

At the onset of puberty, the pituitary gland is stimulated to produce FSH and ICSH by a releaser hormone from the hypothalamus.

HORMONAL CONTROL OF THE MALE REPRODUCTIVE SYSTEM

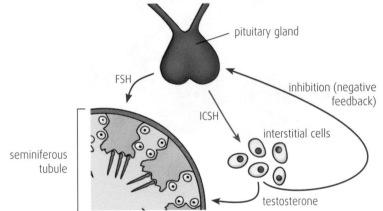

pituitary gland

FSH

inhibition (negative feedback)

ICSH

interstitial cells

seminiferous tubule

testosterone

DON'T FORGET

Hormones reach their target organs through the bloodstream.

Sperm production is under the control of **follicle stimulating hormone (FSH)** and **interstitial cell stimulating hormone (ICSH)**, two hormones that are released from the pituitary gland in the brain; and testosterone released from the testes.

Follicle stimulating hormone (FSH)	FSH acts on the seminiferous tubules to promote sperm production.
Interstitial cell stimulating hormone (ICSH)	ICSH acts on the interstitial cells of the testes, stimulating the production of **testosterone**.
Testosterone	Testosterone is produced by the interstitial cells in the testes. It acts on the seminiferous tubules, stimulating sperm production and activates the prostate and seminal vesicles.

DON'T FORGET

A large number of sperm are needed, as only a tiny fraction will reach the ovum.

Overproduction of testosterone is prevented by a **negative feedback** mechanism. When the testosterone level increases above normal, it inhibits secretion of FSH and ICSH from the pituitary. Testosterone production then stops until the level drops below normal, when the inhibitory effect is switched off and production begins again.

ONLINE TEST

Want to test your knowledge of the male reproductive system? Head to www.brightredbooks.net

THINGS TO DO AND THINK ABOUT

Sterilisation in the male (vasectomy) involves cutting the sperm duct within the scrotum. What effect would this operation have on (i) the level of testosterone in the blood and (ii) sperm production?

FEMALE REPRODUCTIVE SYSTEM

The female displays **cyclical fertility** (eggs are produced only once a month) due to changes in hormone levels throughout the menstrual cycle. This continues from puberty to the menopause.

DON'T FORGET

Ovulation is indicated by an increase in body temperature by about 0·5°C.

MENSTRUAL CYCLE

The menstrual cycle consists of cyclical changes in both the ovary and the uterus.

Changes in the Ovary

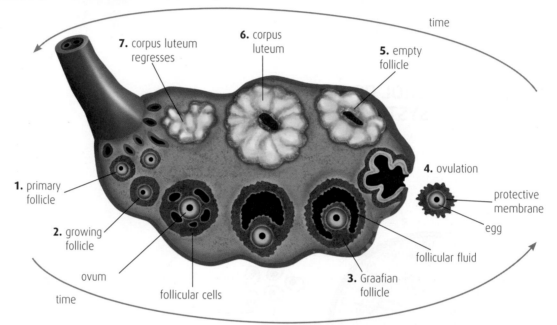

Follicular phase	Under the influence of **follicle stimulating hormone** (**FSH**) from the pituitary gland, **follicles** begin to develop during the **follicular phase**. The follicular cells produce follicular fluid, which gathers within the follicle as it enlarges; and secrete **oestrogen**, which acts on the **pituitary gland**, stimulating the production of **luteinising hormone** (**LH**). Usually only one follicle matures fully to produce a **Graafian follicle**. Peak oestrogen level stimulates an **LH surge**. About 10–12 hours later, the Graafian follicle ruptures (causing **ovulation**) on about day 14 of the menstrual cycle, releasing the ovum.
Luteal phase	The ovary now enters the **luteal phase** where, under the influence of LH, the **corpus luteum** develops from the remaining cells of the ruptured follicle. The corpus luteum produces **progesterone** and **oestrogen**, which have a negative-feedback effect on the pituitary gland, inhibiting the release of FSH and preventing development of any more follicles. If no pregnancy occurs, LH, progesterone and oestrogen levels will decrease and the corpus luteum will degenerate, allowing the development of follicles to begin again in the next cycle. If there is a pregnancy, the corpus luteum continues to function until the placenta is large enough to take over hormone production.

Changes in the Uterus

During the menstrual cycle the following changes occur in the uterus:

Menstruation	Low levels of oestrogen and progesterone cause the lining layer of the uterus (**endometrium**) to be shed. The **menstrual flow** consists of a mixture of endometrial cells, mucus, blood and tissue fluid.

DON'T FORGET

At the onset of puberty, the hypothalamus produces a releaser hormone which stimulates the pituitary gland to secrete FSH, follicle stimulating hormone, and LH, luteinising hormone.

contd

Follicular phase	Increasing oestrogen production from the ovarian follicles stimulates repair of the endometrium. The endometrium becomes thicker and develops a good blood supply. Oestrogen also makes the cervical mucus less viscous, allowing sperm to pass through more easily.
Luteal phase	**After ovulation**, the endometrium continues to thicken under the influence of progesterone from the corpus luteum. This is accompanied by further development of the endometrial blood vessels and endometrial glands, preparing the uterus for implantation of the blastocyst, should fertilisation take place. If there is no fertilisation, the drop in progesterone level that results from degeneration of the corpus luteum causes the uterus to enter the next menstrual phase. If fertilisation does occur, the endometrium will be maintained.

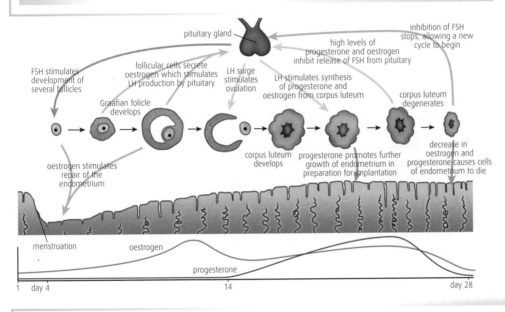

ONLINE

Test your knowledge of the female reproductive system at www.brightredbooks.net

ONLINE

Learn more about ovarian and uterine cycles by following the link at www.brightredbooks.net

FERTILISATION AND IMPLANTATION

If gametes are viable, the following sequence of events can occur after intercourse:

Fertilisation	Sperm usually reach the ovum in the **oviduct**. Here, they release enzymes that enable them to penetrate the layer surrounding the ovum, allowing fertilisation to occur. The fertilised ovum is called a **zygote**. To prevent fertilisation by more than one sperm, the ovum changes the chemical structure of the layers that surround it, making it impossible for other sperm to pass through.
Cleavage	As the zygote travels down the oviduct towards the uterus, it divides by a process called **cleavage** to produce a ball of cells called a blastocyst.
Implantation	After about 7 days, implantation takes place, when the blastocyst enters the uterus and embeds in the endometrium.
Differentiation	Embryonic cells now differentiate to produce specialised tissues of the embryo, embryonic components of the placenta and the amniotic sac.

 THINGS TO DO AND THINK ABOUT

The menstrual cycle is under the control of hormones released by both the ovary and the pituitary gland. Make a list of the effects of each of the following hormones on the organs of the female reproductive system:

1 Follicle stimulating hormone
2 Luteinising hormone
3 Oestrogen
4 Progesterone.

 DON'T FORGET

If a pregnancy results, the placenta will take over production of progesterone, maintaining the endometrium and preventing miscarriage.

CONTRACEPTION AND INFERTILITY

ONLINE

For an NHS guide to contraception, see: www.nhs.uk/Conditions/Contraception

BIOLOGICAL BASIS OF CONTRACEPTION

Contraceptive methods act by preventing either fertilisation or implantation. You should, in particular, be familiar with the use of hormonal methods and determination of the fertile period.

Hormonal Methods

The contraceptive pill contains a combination of the hormones oestrogen and progesterone. These mimic negative feedback by inhibiting production of FSH and LH from the pituitary gland. As a result, follicles in the ovary do not mature and ovulation does not take place.

Determination of the Fertile Period

Females may try to prevent conception by avoiding intercourse around the fertile period, when conception is possible. As ova survive for approximately 24 hours after ovulation and sperm can live for about 3 days in the female reproductive tract, the fertile period starts about 3 days before ovulation and ends the day after ovulation.

In the example shown, the menstrual cycle lasts 28 days, with ovulation occurring on day 14. The fertile period starts on day 11, as sperm deposited in the female reproductive tract on this day would still be capable of fertilising the ovum on day 14. Day 15 is the last day of the fertile period as the ovum would not be viable after this day.

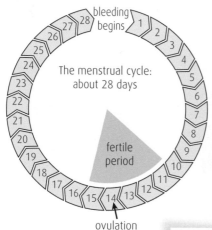

The menstrual cycle: about 28 days

bleeding begins

fertile period

ovulation

INFERTILITY AND ITS TREATMENT

For fertilisation and implantation to occur, viable gametes must be produced and the following events must be both possible and coordinated:

- The ovum must be able to travel down the uterine tube.
- Sperm must be able to swim through the female reproductive tract to reach and fertilise the ovum.
- The endometrium must be ready to receive the embryo.

About 10% of couples are infertile, with the most common cause being failure to ovulate. You should be familiar with the different causes of infertility and methods that can be used to treat them.

Failure to ovulate	This is usually the result of a **hormone imbalance** and is treated by using **fertility drugs** that mimic the action of FSH and LH from the pituitary gland. Drinking excessive quantities of alcohol, smoking, stress and obesity can be risk factors.
Low sperm count	A low level of sperm production is often caused by a **hormone imbalance** and can be treated using **testosterone**, *in vitro fertilisation*, or **artificial insemination** using donor sperm. Additional risk factors include smoking, stress and excessive alcohol consumption.
Blockage of uterine tubes	Uterine tubes can become blocked through **infection** or abnormal tissue growth. Where blockages cannot be surgically removed, *in vitro fertilisation* may be an option.
Failure to implant	For implantation to occur, the monthly changes in the endometrium must be synchronised with changes in the ovary, so that the endometrium is thick enough to receive an embryo as it enters the uterus. **Hormone imbalances** can prevent coordination of the ovarian and uterine changes and can be treated using **fertility drugs**.

ONLINE

Learn more about infertility and the various treatments available at www.brightredbooks.net

contd

In Vitro Fertilisation

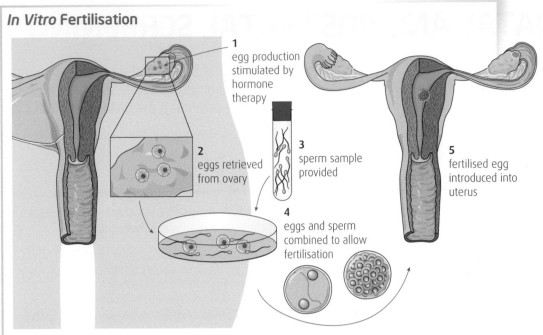

1 egg production stimulated by hormone therapy

2 eggs retrieved from ovary

3 sperm sample provided

4 eggs and sperm combined to allow fertilisation

5 fertilised egg introduced into uterus

This technique involves giving hormones to the female, causing several ova to be released at ovulation. A syringe is inserted into the female's abdominal cavity, allowing the ova to be collected and placed in liquid nutrient. Sperm are either added to the liquid or injected into an ovum, bringing about fertilisation. Several embryos are then inserted into the female's uterus.

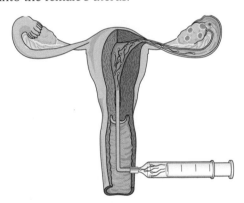

Artificial Insemination

In artificial insemination, semen is collected and inserted into the female reproductive tract using a catheter, without intercourse having taken place. If a male has a low sperm count, multiple samples of his sperm can be combined and then inserted into his partner's reproductive tract to increase the chances of fertilisation. If the male is infertile, sperm from a donor can be used to inseminate the female.

Intracytoplasmic Sperm Injection (ICSI)

Sperm are collected either after ejaculation or, if this is not possible, directly from the epididymis or testes. A single sperm is selected and the head is injected directly into an egg. If fertilisation is successful, between one and three embryos will be selected for transfer into the female's uterus.

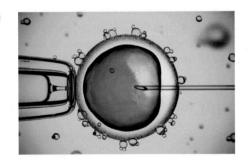

 THINGS TO DO AND THINK ABOUT

Barrier methods of contraception

Barrier methods of contraception work by placing a physical barrier between the sperm and ovum, preventing fertilisation from occurring. Spermicides (chemicals that kill sperm) are often used at the same time as the barrier method, in order to increase its effectiveness. Research the common barrier methods of contraception, such as condoms, cervical cap and diaphragm.

 ONLINE TEST

Head to www. brightredbooks.net and test yourself on contraception and infertility.

ANTENATAL AND POSTNATAL SCREENING

At antenatal appointments, a pregnant woman is asked health and family history questions, and undergoes **screening tests**. This allows calculation of the risk of some disorders, so that – if any problems are indicated – she can be offered **diagnostic tests** to provide a **prenatal diagnosis**. A **genetic counsellor** will discuss any diagnosis with the family.

DON'T FORGET

Screening tests are routinely made for thalassaemia, Down's syndrome and spina bifida, but can also include cystic fibrosis, sickle-cell anaemia and Tay Sachs disease.

DON'T FORGET

False positive tests can result if substances are measured at the wrong stage of pregnancy.

DON'T FORGET

Antibodies cross the placenta from mother to foetus.

ANTENATAL SCREENING TESTS

Maternal and foetal health **screening tests** include measuring maternal weight, height and blood pressure, as well as monitoring changes in particular chemicals.

Weight and height	Weight and height are measured to calculate body mass index (BMI) as women who are overweight are at greater risk of problems during pregnancy, including gestational diabetes, miscarriage, high blood pressure and pre-eclampsia, and are also more likely to require either an instrumental delivery or a caesarean section.
Blood pressure	High blood pressure (hypertension) affects between 10–15% of pregnancies in the UK. In later pregnancy this can be an indicator of pre-eclampsia.

Biochemical Testing

Biochemical tests are used to chart the normal physiological changes that take place during pregnancy.

Urine is tested for protein. This is an indicator of infection, diabetes or pre-eclampsia.

Blood tests include:
- Rhesus-antibody testing – females who are Rhesus negative and who have not developed anti-D antibodies will be offered anti-D injections at 28 and 34 weeks of pregnancy and also after the birth.
- Anaemia – if anaemic, the female is offered iron and folic acid supplements.

Ultrasound Imaging

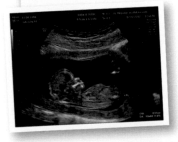

In the UK, two ultrasound scans are usually offered. A **dating scan** is made at between 8 and 12 weeks, to allow the gestational age and due date to be determined. An **anomaly scan** is made at between 18 and 20 weeks to detect any physical problems.

Images are produced when high-frequency sound waves travel from a probe (transducer), through gel on the skin, and then bounce back from body organs. The great benefit of ultrasound images is that they can show, not only the structure of the baby's organs, but also their movements and blood flow.

DIAGNOSTIC TESTING

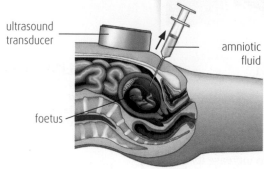

ultrasound transducer

amniotic fluid

foetus

If screening tests, family history or maternal risk category indicate the need for further testing, parents are offered diagnostic testing.

Amniocentesis

A needle is inserted through the wall of the abdomen and uterus, into the amniotic sac, to collect a sample of amniotic fluid. This contains foetal skin and hair cells. The extracted cells are grown in tissue culture before their chromosomes are examined in a karyotype.

Anmiocentesis has a low risk of miscarriage (0·5–1% above normal) but cannot be performed until between 16 and 18 weeks of pregnancy, when the mother is beginning to feel an attachment to her unborn child.

contd

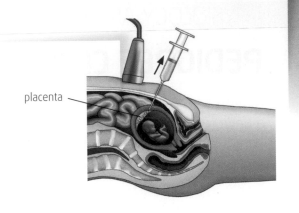

placenta

Chorionic Villus Sampling

A small sample of foetal cells from the placenta is removed, either through the wall of the abdomen or through the vagina. The extracted cells are grown in tissue culture and their chromosomes are examined in a karyotype.

This test can be carried out at between 8 to 10 weeks. As part of the placenta is being removed, the risk of miscarriage is increased (2–3% above normal). Also, maternal cells may be extracted with the foetal cells, giving a false result.

Karyotyping

Once foetal cells have been obtained, they are grown in tissue culture. Pictures of the foetal chromosomes are taken during mitosis, and a **karyotype** is constructed. The complete foetal chromosome complement is analysed by arranging the chromosomes in order by **size**, **shape** and **banding pattern**. A genetic counsellor looks for missing, extra or abnormal chromosomes.

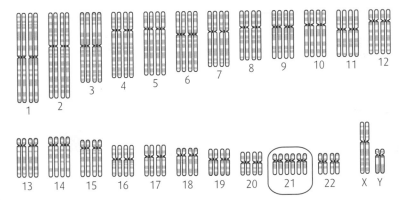

The karyotype here shows Down's syndrome, where there is an extra chromosome number 21.

PGD in IVF

During *in vitro* fertilisation (IVF) procedures, it is possible to extract embryonic cells for genetic profiling prior to implantation. Pre-implantation genetic diagnosis (PGD) allows embryos free of specific genetic conditions to be selected for implantation.

POSTNATAL SCREENING

After birth, it is possible to carry out biochemical tests for several inherited conditions, including phenylketonuria (PKU). Routine screening for PKU is currently undertaken in the UK after every birth. If it is not detected and treated very quickly after birth, postnatal development of the brain is affected. Affected individuals must be given a reduced phenylalanine diet.

THINGS TO DO AND THINK ABOUT

Nuchal Translucency (NT) Scan

This is a screening test which estimates the risk of Down's syndrome and other chromosomal abnormalities using an ultrasound scan at between 11 and 13 weeks plus 6 days, when the foetus is between 45 mm and 84 mm long. The amount of fluid under the skin at the back of the neck (the nuchal translucency) is measured and used, along with maternal age, to determine risk. 75% of babies with Down's syndrome can be detected and the accuracy can be increased by using a combined scan and blood test. If Down's syndrome is indicated, amniocentesis or chorionic villus sampling is offered.

Rhesus Factor

The placenta acts as an immune barrier, preventing the mother's body from recognising foetal antigens as foreign. However, should maternal and foetal blood mix during miscarriage or birth, the mother's immune system produces antibodies. If the blood of a Rhesus negative (Rh^-) mother mixes with blood from a Rhesus positive (Rh^+) baby during birth, the mother's immune system produces anti-D antibodies and memory cells. During subsequent pregnancies, a Rh^+ foetus will be attacked by anti-D antibodies. To prevent this immune response, a Rh^- mother can be given an injection of anti-D antibodies after the birth of the first baby, destroying any of the baby's Rh^+ blood cells in the mother's blood.

ONLINE

Head to www.brightredbooks.net to see a pregnancy and new-born screening timeline.

ONLINE TEST

How well have you learned about antenatal and postnatal screening? Take the test at www.brightredbooks.net

PEDIGREE CHARTS (FAMILY TREES)

Genetic counsellors discuss the family history of inherited disorders with patients. This allows a pedigree chart to be constructed. From the chart, the counsellor determines the pattern of inheritance and calculates the risk of a condition being passed on to the next generation.

The key below should be used with each of the pedigree charts.

Key

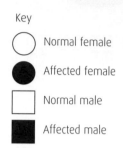

○ Normal female

● Affected female

□ Normal male

■ Affected male

AUTOSOMAL DOMINANT INHERITANCE

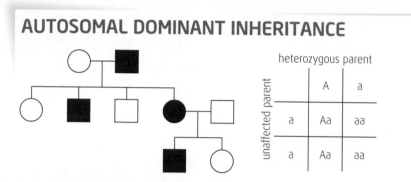

	heterozygous parent	
unaffected parent	A	a
a	Aa	aa
a	Aa	aa

- Parents who are unaffected have unaffected children.
- Affected individuals with unaffected partners have a 50% chance of having an affected child (see punnet square above).
- The characteristic is seen in every generation.
- Both sexes are equally affected.

Examples of autosomal dominant conditions are **Huntington's chorea** and **achondroplasia**.

ONLINE

Learn more about autosomal dominant and recessive conditions by following the links at www.brightredbooks.net

AUTOSOMAL RECESSIVE INHERITANCE

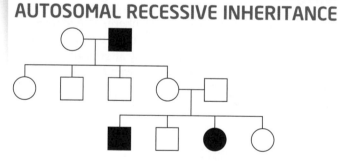

	affected parent	
carrier	a	a
A	Aa	Aa
a	aa	aa

	carrier	
carrier	A	a
A	AA	Aa
a	Aa	aa

- Parents who are both heterozygous can have affected children.
- Children of affected parents always inherit the recessive allele.
- Both sexes are equally affected.
- The condition can skip generations.

Examples of autosomal recessive conditions are **cystic fibrosis** and **PKU**.

SEX-LINKED INHERITANCE

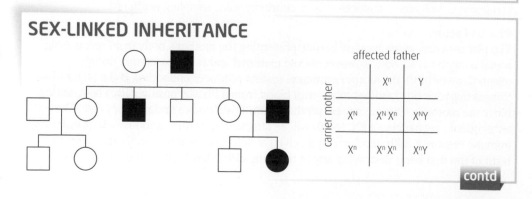

	affected father	
carrier mother	X^n	Y
X^N	$X^N X^n$	$X^N Y$
X^n	$X^n X^n$	$X^n Y$

contd

- More males than females are affected.
- Only females can be carriers.
- Daughters of affected males are either carriers or affected.
- A carrier with a normal partner has a 50% chance of having an affected son or a carrier daughter.
- The condition can skip generations.

The punnet square shows the possible genotypes of children whose parents were a carrier female and affected male.

Examples of sex-linked inheritance are **Duchenne muscular dystrophy**, **haemophilia** and **red–green colour blindness**.

DON'T FORGET

Sex-linked genes are always indicated using the letters X and Y in the genotype.

INCOMPLETE DOMINANCE INHERITANCE

Where two alleles are **incompletely dominant**, a heterozygous individual will have a phenotype that is a blend of the two homozygous types. An example of incomplete dominance is sickle-cell trait. The gene which codes for haemoglobin has two alleles: allele A codes for normal haemoglobin and allele S codes for haemoglobin S. Unaffected individuals have a genotype AA, and their red blood cells contain normal haemoglobin. The table below shows the other genotypes and phenotypes associated with this gene.

Genotype	Phenotype	Description of condition
SS	sickle-cell anaemia	Individuals who are homozygous for the haemoglobin S allele suffer from sickle-cell anaemia. Haemoglobin S binds with less oxygen than normal haemoglobin. Red blood cells are sickle shaped and clump together, which blocks blood vessels. People with this condition often die at a young age.
AS	sickle-cell trait	Heterozygous individuals suffer a less serious condition called sickle-cell trait. Red blood cells are a normal biconcave shape and contain both normal haemoglobin and haemoglobin S.

In the following example, a woman who has sickle-cell trait and a man who is unaffected have children together. The expected frequency of each phenotype in the first generation is shown.

		mother		father
parents				
phenotype		sickle-cell trait	×	unaffected
genotype		AS		AA
gametes		A or S		all A

F₁ punnet square

	A	A
A	AA	AA
S	AS	AS

Genotype	Phenotype	Expected frequency
AA	unaffected	50%
AS	sickle-cell trait	50%

THINGS TO DO AND THINK ABOUT

Risk evaluation in polygenic inheritance
The disorders described above are all single-gene disorders. Polygenic disorders are caused by more than one gene working in combination. You should be aware that it is more difficult to accurately measure the probability of inheriting a genetic disorder when the condition is polygenic. This is because environmental factors, as well as genes, play a role in determining the severity of the condition. Examples of polygenic disorders include diabetes and asthma.

ONLINE TEST

Test yourself on pedigree charts at www.brightredbooks.net

CARDIOVASCULAR SYSTEM

The cardiovascular system is composed of:

- Blood – consisting of red and white blood cells, platelets and plasma.
  The blood transports oxygen and carbon dioxide, food molecules and hormones around the body. It also carries heat around the body. (The liver is responsible for generating most heat when the body is at rest.)
- Blood vessels – which transport the blood around the body.
- The heart – which pumps blood through the vessels.

artery

capillary

wall one cell thick

vein

BLOOD VESSELS

The direction of blood flow through the circulatory system is as follows:

Heart

Artery

Arteriole

Capillary

Venule

Vein

The heart is the muscular pump which pushes blood around the system.

Arteries carry blood away from the heart. They have thick walls made of: an outer layer of connective tissue, which contains elastic fibres; a middle layer of smooth muscle and elastic fibres; an inner lining layer of cells, called the endothelium. Elastic fibres allow the vessel to stretch as blood pulses through. This is what you feel when you take your pulse. The thick muscle layer allows the artery to withstand the high pressure produced by the heart muscle.

Arterioles are the terminal branches of arteries and lie in the body tissues. Their walls contain a lot of smooth muscle, which can contract to narrow the blood vessel.

Capillaries are the smallest blood vessels and have very thin walls, just one cell thick. As blood flows through a capillary, liquid is forced through the walls by the blood pressure and enters the tissue fluid. The capillaries are so narrow that red blood cells must squeeze through, slowing down the rate of blood flow, giving more time for exchange of materials, and reducing the blood pressure. The blood pressure is, therefore, higher at the arterial end and lower at the venous end where some of the tissue fluid drains back into the capillary to be returned to the heart.

Blood leaving the capillaries enters a series of thin-walled venules.

Veins carry blood back towards the heart. An endothelium lines the lumen of the vessel. Outside this, lies a much thinner layer of smooth muscle than is found in arteries of similar size (a thicker layer is not needed as the blood is at low pressure). Connective tissue containing elastic fibres forms the outermost layer. Valves are present in the walls to prevent the backflow of blood. Blood flow through the veins is assisted by contraction of muscles in the surrounding tissues, for example in the leg, where contraction of the calf muscles helps to push blood back up the lower limb.

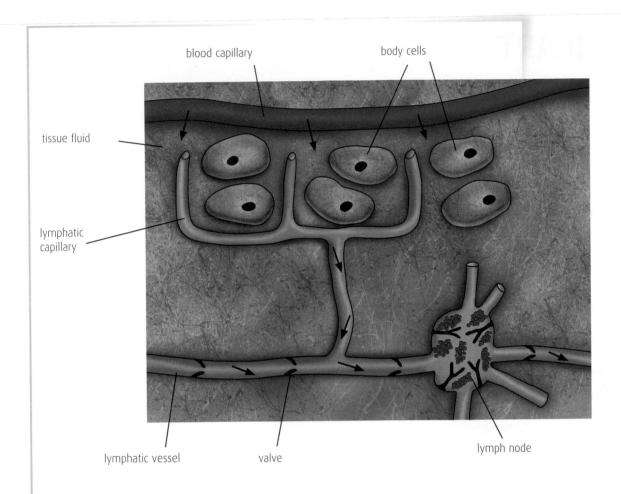

blood capillary

body cells

tissue fluid

lymphatic capillary

lymphatic vessel

valve

lymph node

LYMPHATIC SYSTEM

Most cells in the human body are not in contact with the blood. Instead they are bathed in tissue fluid, which has passed out through the walls of the blood capillaries. Tissue fluid allows exchange of metabolites between tissue cells and the blood. Some tissue fluid drains back into the bloodstream, but most will enter blind-ending tubes called **lymphatic capillaries** to form **lymph**. Lymphatic capillaries drain into larger **lymphatic vessels**, which link up to form the lymphatic system. As there is no pump in the lymphatic system (unlike the blood transport system which has the heart), flow of lymph is brought about by contraction of muscles in the surrounding tissues. Valves within the lymphatic vessels prevent backflow of fluid.

The largest lymphatic vessels return lymph to the blood circulatory system by draining into two large veins in the chest (the subclavian veins). As lymph flows through the lymphatic vessels, it passes through swellings called **lymph nodes**. These can be found either singly or accumulated in groups in areas such as the groin and the armpit.

 THINGS TO DO AND THINK ABOUT

1 Explain the function of (a) elastic fibres in artery walls and (b) smooth muscle fibres in arteriole walls.
2 Which blood vessels have valves in their walls? What is their function?
3 In what way are tissue fluid and blood plasma different?

 **DON'T FORGET**

Materials are exchanged between tissue fluid and body cells by pressure filtration.

 VIDEO LINK

Learn more about the cardiovascular system by watching the clip at www.brightredbooks.net

 ONLINE TEST

Head to www.brightredbooks.net and test yourself on the cardiovascular system.

THE HEART

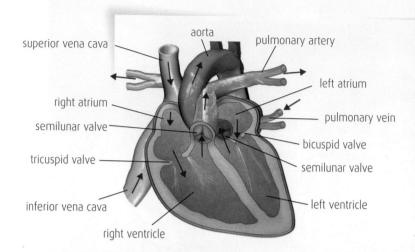

aorta

superior vena cava

pulmonary artery

left atrium

right atrium

pulmonary vein

semilunar valve

bicuspid valve

tricuspid valve

semilunar valve

inferior vena cava

left ventricle

right ventricle

INTRODUCTION

The heart is the muscular organ which pumps blood around the body. It is often referred to as a double pump: the right side of the heart pumps deoxygenated blood to the lungs and the left side of the heart pumps oxygenated blood to all parts of the body. The heart is made up of four chambers: two **atria** that receive blood from the main veins and two **ventricles** that pump blood either to the lungs (right ventricle) or to the body (left ventricle). The heart muscle (cardiac muscle) is supplied by the **coronary arteries**. Valves within the heart are present to prevent backflow of blood.

Name of valve	Location	Phase of cardiac cycle when valve is closed	Function of valve
Atrioventricular valves (tricuspid and bicuspid)	Between the atria and ventricles	Ventricular systole	Prevent the backflow of blood into the atria
Semilunar valves	At the start of the pulmonary artery (on the right) and the aorta (on the left)	Atrial systole	Prevent the backflow of blood from the main arteries into the ventricles

CARDIAC CYCLE

The sequence of filing and emptying of the heart chambers is called the **cardiac cycle**. During the cardiac cycle, contraction and relaxation of cardiac muscle alters the blood pressure within each of the heart chambers, causing the correct flow of blood through the heart. Blood will always flow from high to low blood pressure unless a valve is closed, preventing blood flow. The cardiac cycle is divided into periods of relaxation (**diastole**) and periods of contraction (**systole**).

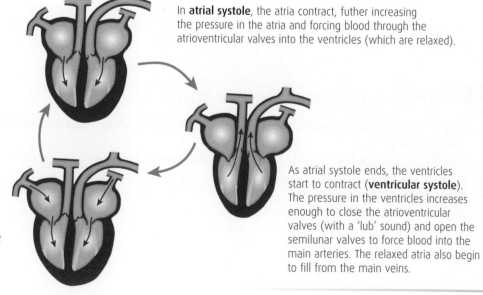

In **atrial systole**, the atria contract, futher increasing the pressure in the atria and forcing blood through the atrioventricular valves into the ventricles (which are relaxed).

In **diastole**, the ventricles relax, causing the pressure to drop below that of the main arteries, closing the semilunar valves (with a 'dup' sound). The atria are relaxed and continue to fill with blood from the vena cava and pulmonary vein, increasing the pressure above that of the ventricles. This forces the atrioventricular valves open, and the ventricles begin to fill.

As atrial systole ends, the ventricles start to contract (**ventricular systole**). The pressure in the ventricles increases enough to close the atrioventricular valves (with a 'lub' sound) and open the semilunar valves to force blood into the main arteries. The relaxed atria also begin to fill from the main veins.

CONTROL OF THE CARDIAC CYCLE

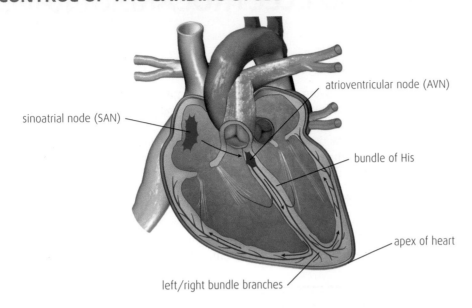

sinoatrial node (SAN)

atrioventricular node (AVN)

bundle of His

apex of heart

left/right bundle branches

Although cardiac muscle is able to beat on its own, the contraction of each heart chamber must be coordinated to bring about the correct movement of blood.

Coordination of the cardiac cycle is brought about by the **conducting system** of the heart. Electrical excitement (a **cardiac impulse**) is initiated in an area of the right atrium called the **sinoatrial node (SAN)**, the pacemaker of the heart, where the cells are **auto-rhythmic**. From here, a wave of contraction moves out across the atria to reach the **atrioventricular node (AVN)** in the right atrium. The impulse passes down through a bundle of fibres in the central wall of the heart to reach the apex of the heart, then up through left and right branches to the walls of the ventricles. Ventricular contraction (systole) begins at the apex of the ventricles and spreads upwards to squeeze blood out of the ventricles towards the main arteries.

The heart rate is under both nervous and hormonal control.

- **Sympathetic nerves** increase the heart rate.

- **A parasympathetic nerve**, the vagus nerve, slows down the heart rate.

- The part of the brain controlling these nerves is called the **medulla oblongata**.

- **Noradrenaline (nor-epinephrine)** is released by the sympathetic accelerator nerves and speeds up the heart rate, and **acetylcholine** – released by parasympathetic nerves – slows the heart rate.

 THINGS TO DO AND THINK ABOUT

Which line in the table correctly identifies the state of the heart valves during ventricular systole?

	Semi-lunar valves	Atrio-ventricular valves
A	open	open
B	open	closed
C	closed	open
D	closed	closed

ONLINE

For more on the human heart, follow the link at www.brightredbooks.net

DON'T FORGET

To follow the cardiac cycle, you only need to consider one side of the heart (as the right side is always at the same stage as the left side).

 ONLINE TEST

Test your knowledge of the heart at www.brightredbooks.net

BLOOD PRESSURE AND ELECTROCARDIOGRAMS

DON'T FORGET

Hypertension is a major risk factor for coronary heart disease.

ONLINE

For more on the cardiac cycle, follow the link at www.brightredbooks.net

BLOOD PRESSURE

Blood pressure is measured using a sphygmomanometer. A cuff is placed around the upper arm and inflated to temporarily stop blood flow through the artery. For manual devices, a stethoscope is used to listen for sounds in the artery as the pressure is slowly released. This allows the practitioner to determine when the pulse is first heard (**systolic pressure**) and then when the sounds stop again (**diastolic pressure**). A typical blood pressure reading for a young adult is 120/70 mm Hg.

Blood Pressure in the Heart

You should remember that blood will always flow from high to low pressure, unless a closed valve prevents this. The diagram below shows the changes in pressure that occur during the cardiac cycle in the left atrium, left ventricle and the aorta.

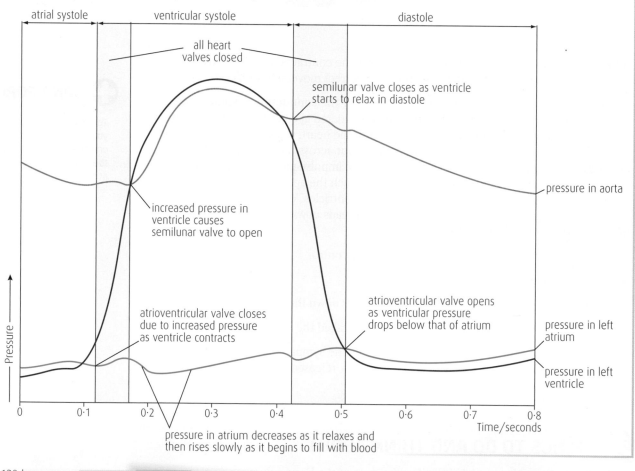

BLOOD PRESSURE IN THE BLOOD VESSEL

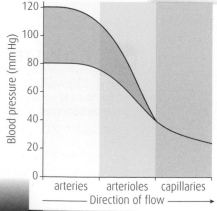

In the vessels, blood pressure is caused by the pumping action during contraction and relaxation of the heart ventricles. During ventricular systole, blood pressure is at its highest and it decreases during ventricular diastole. In the arteries, the walls bulge during systole as a wave of blood passes through and recoil during diastole, pushing blood through the vessel. As blood enters narrower vessels, the resistance of the vessel wall increases. This slows blood flow, reducing blood pressure as blood continues travelling along the vessel. The largest decrease in blood pressure takes place in the arterioles, where resistance is at its highest. Blood pressure continues to decrease as blood flows through capillaries, venules and veins.

INTERPRETATIONS OF ELECTROCARDIOGRAMS (ECG)

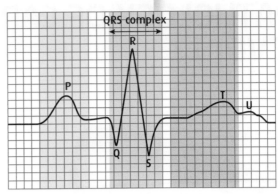

An electrocardiogram records the electrical impulses that are generated in the heart during contraction and relaxation of the cardiac muscle. It is shown on graph paper as a wave. Abnormal ECGs can indicate various abnormalities, including damage due to a heart attack, abnormal rhythms (arrhythmia) and high blood pressure.

- The P wave is caused by firing of the sinoatrial node, with a wave of electrical activity crossing the atria causing atrial contraction.

- Firing of the atrioventricular node produces the QRS complex and initiates the contraction of the ventricles, with electrical activity spreading through the ventricles.

- The T wave results from the AV node recovering. The ventricles return to their resting electrical state.

DON'T FORGET

The volume of blood pumped by left and right ventricles is equal.

CARDIAC OUTPUT

The **cardiac output** (CO) is the volume of blood pumped through each ventricle per minute. To calculate this, two values must be known:

- **heart rate** (HR) – the number of contractions in one minute

- **stroke volume** (SV) – the volume of blood pumped out by one ventricle during systole.

The following equation is used to calculate cardiac output:

$$CO = HR \times SV$$

For example, if an individual has a heart rate of 70 beats per minute and a stroke volume of 0·08 litres, then:

$$CO = 70 \times 0·08 = 5·6 \text{ l/min}$$

THINGS TO DO AND THINK ABOUT

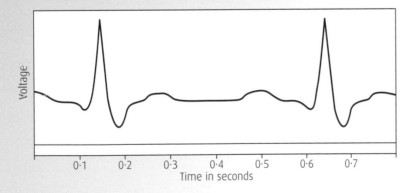

What heart rate is shown on the electrocardiogram (ECG)?

Find the time taken for one heartbeat by identifying the time between two identical points on the graph, e.g. 0·1 and 0·6 seconds; so, it takes 0·5 seconds for one heartbeat. Every second there will be 2 heartbeats, giving a heart rate of:

$$2 \times 60 = 120 \text{ beats per minute}$$

ONLINE TEST

Revise your knowledge of blood pressure and electrocardiograms by taking the topic test at www.brightredbooks.net

PATHOLOGY OF CARDIOVASCULAR DISEASE (CVD) 1

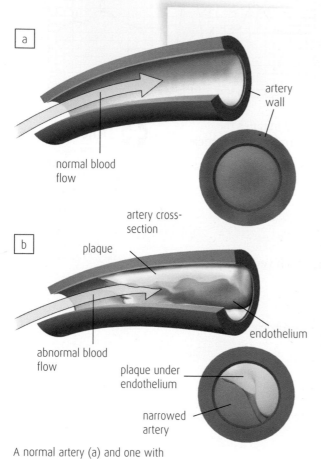

normal blood flow

artery wall

artery cross-section

plaque

abnormal blood flow

plaque under endothelium

endothelium

narrowed artery

A normal artery (a) and one with atherosclerosis (b)

ATHEROSCLEROSIS

In **atherosclerosis** the normally smooth wall of the blood vessel (diagram a) becomes roughened by the development of an **atheroma** or **plaque** below the endothelium (diagram b). The atheroma is composed of fibrous material, calcium and fatty materials, including cholesterol. The atheroma gradually gets bigger, causing the artery wall to thicken and the diameter of the lumen to decrease. As a result, less blood flows through the vessel – eventually it may block completely. Blood pressure also increases.

Risk factors which can make individuals more likely to develop an atheroma include lack of exercise, a high fat diet, obesity, diabetes and high blood pressure.

Thrombosis

Thrombosis is the formation of a clot (**thrombus**) within a blood vessel. This can be initiated by rupture of an atheroma, causing damage to the endothelium. When this happens, clotting factors set off a chain (a cascade) of linked reactions that convert the inactive enzyme **prothrombin** to the active enzyme, **thrombin**. Part of the cascade is shown below.

Thrombin is the enzyme which catalyses the conversion of the soluble plasma protein **fibrinogen** into insoluble threads of **fibrin**. As the threads are formed, a mesh builds up which traps blood cells and platelets, forming a **clot**. Sometimes, the thrombus breaks free from its site of formation (forming an **embolus**) and travels around in the blood until it lodges in a

DON'T FORGET

Atherosclerosis is the major risk factor for many cardiovascular diseases, including angina, heart attack, stroke and peripheral vascular disease.

ONLINE

Learn more about atherosclerosis by following the link at www.brightredbooks.net

```
                        ┌──────────────┐
                        │ vessel damage│
                        └──────┬───────┘
                               │
                               ▼
    inactive            active
    clotting factors ──▶ clotting factors
                               │
                               ▼
    prothrombin ──────────▶ thrombin
                               │
                               ▼
    fibrinogen ──────────▶ fibrin
```

blocking blood flow. If either a thrombus or embolus blocks blood flow in an artery, the tissue that the vessel supplies is deprived of oxygen and the cells will die. Blockage of the coronary artery can cause a heart attack (myocardial infarction, **MI**); blockage of arteries in the brain can cause a **stroke**.

Peripheral Vascular Disorders

Where atherosclerosis affects arteries other than those of the brain and heart, the disease is referred to as a **peripheral vascular disorder**. A progressive narrowing of the arteries, most often in the legs, reduces the oxygen supply to the muscles. During exercise, the muscle cells cannot obtain enough oxygen, resulting in pain in the calves and thighs. The poor blood flow can also cause cold and painful fingers and toes.

Deep vein thrombosis (DVT) results from the production of a blood clot in one of the deep veins. This most frequently occurs in the legs, where slowed blood circulation during long periods of inactivity – either through sitting down (in which case the veins behind the knee joint become kinked) or lying in bed – can add to the effects of atherosclerosis. If an embolus breaks free from the DVT, it can travel to the lung, causing a blockage to one of the pulmonary vessels (a **pulmonary embolism**).

ONLINE TEST

Test yourself on the pathology of cardiovascular disease at www. brightredbooks.net

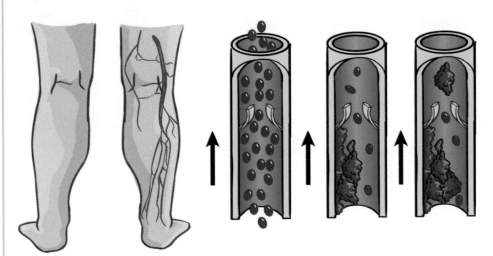

Deep veins of the leg   Normal blood flow   Deep vein thrombosis   Embolus

 THINGS TO DO AND THINK ABOUT

Thrombolytic medications
Some 'clot busting' drugs, such as **streptokinase** and **tissue plasminogen activator**, can be used for the immediate treatment of stroke, heart attack, DVT and pulmonary embolism. Plasminogen is activated by these drugs, producing plasmin which dissolves fibrin and makes the blood clot soluble.

Using your knowledge of the pathology of cardiovascular diseases, make a bullet point list of 10 important facts on the occurence of cardiovascular disease.

PATHOLOGY OF CARDIOVASCULAR DISEASE (CVD) 2

ONLINE

Learn more about high cholesterol at www. brightredbooks.net

CHOLESTEROL

Cholesterol is a vital substance in the body. It is used in the production of steroid hormones, such as testosterone and oestrogen, and is also a component of cell membranes. About 25% of cholesterol is produced in the liver, but all cells are capable of its synthesis. About 15% is obtained in the diet from the saturated fats found in meat, dairy products and egg yolk. A diet high in saturated fats raises the blood cholesterol level. This, in itself, is not a disease but can lead to the development of atherosclerosis.

Lipoproteins

Cholesterol is transported in the blood by molecules called lipoproteins. These are made up of an outer layer consisting of phospholipids and some proteins. Lying in the centre of the lipoprotein are fats and steroids, including cholesterol.

The two main groups of lipoproteins are:

protein

triglycerides and cholesterol

phospholipid monolayer

- **High-density lipoprotein (HDL)** – this transports cholesterol from body cells to the liver, where it is eliminated by breakdown into useful chemicals such as bile salts. A high HDL level can reduce the blood cholesterol level.

- **Low-density lipoprotein (LDL)** – this transports cholesterol from the liver to body cells. A high LDL level can increase the blood cholesterol level.

Role of Lipoprotein in CVD

The extracellular side of the cell membrane contains LDL receptors. When LDL binds to receptors, endocytosis takes place; LDL enters the cell within a vesicle. The vesicle binds to lysosomes and LDL is broken down to produce amino acids and cholesterol. These are released into the cytoplasm of the cell.

Cholesterol, once released into the cytoplasm, suppresses the enzyme which controls the rate of cholesterol biosynthesis, turning off cholesterol synthesis in the cell. It also turns off the synthesis of LDL receptors and prevents any more LDL from entering the cell. As the cell can no longer take in LDL, this circulates in the blood. Cholesterol may be deposited in the arteries, adding to atheromas.

The ratio of HDL to LDL is extremely important in influencing the occurrence of atherosclerosis:

- A low LDL:high HDL ratio results in a lower blood cholesterol level, reducing the chance of developing atherosclerosis.

- A high LDL:low HDL ratio results in a higher blood cholesterol level, increasing the chance of developing atherosclerosis.

Protective Measures

Several lifestyle choices can lead to a high LDL:low HDL ratio and, therefore, increase the cholesterol level in the blood. Smoking, obesity, lack of exercise, and a diet high in saturated fats are all risk factors. By modifying your lifestyle through regular exercise and not smoking, the level of HDL in the blood can be increased. Replacing saturated animal fats in the diet with unsaturated fats also helps to decrease the level of LDL in the blood.

contd

If altering diet and taking exercise do not reduce cholesterol level, drugs such as statins can be prescribed. Statins work by blocking the enzyme in liver cells that catalyses the formation of cholesterol. As the liver cells now cannot make their own cholesterol, they increase the number of LDL receptors on their cell membranes, resulting in more LDL being taken into the cell and reducing the blood cholesterol level. Patients are advised to take statins in the evening because this enzyme is more active at night.

HYPERCHOLESTEROLAEMIA

Familial hypercholesterolaemia (FH) has an autosomal dominant pattern of inheritance. The most common form of FH is heterozygous (about 1 in every 500 people), with individuals having one mutated and one unaltered allele. Heterozygous individuals have a predisposition to develop very high blood cholesterol level, usually from birth. Individuals who are homozygous are very rare (about 1 in a million) but have a much more severe form of FH with exceptionally high blood cholesterol level and a risk of developing heart disease in childhood.

Mutations of genes involved in FH cause changes in cholesterol metabolism by:

- decreasing the number of LDL receptors on liver cell membranes

or

- changing the structure of the LDL receptor so that LDL molecules cannot bind to it.

Genetic Screening for FH

Diagnosis of FH involves clinical testing and genetic screening. Genetic screening through DNA sequencing has identified over 1000 different mutations on FH genes, with a substitution mutation being the most frequent alteration to the LDL receptor gene. Once a mutation has been identified in a patient, other family members can be tested and a family tree constructed (a pedigree analysis).

Treatment

For heterozygous FH, improved lifestyle and use of clot-busting drugs, such as statins, can reduce the LDL level. For homozygous individuals, statin and non-statin cholesterol-lowering drugs must be used. In the most severe cases, a liver transplant may be necessary.

THINGS TO DO AND THINK ABOUT

In the table, identify the correct relationship between the ratio of HDL:LDL in the blood, blood cholesterol level and the risk of atherosclerosis.

	Cholesterol level	HDL:LDL ratio	Chance of atherosclerosis
A	Low	Low	Increased
B	Low	High	Lowered
C	High	Low	Lowered
D	High	High	Increased

DON'T FORGET

In autosomal dominant conditions such as FH, the condition does not skip generations.

ONLINE TEST

Head to www.brightredbooks.net to test yourself on the pathology of cardiovascular disease.

BLOOD GLUCOSE LEVEL AND DIABETES

REGULATION OF BLOOD GLUCOSE LEVEL

Carbohydrates obtained in the diet are broken down into glucose. Glucose is used by body cells as the main respiratory substrate and a constant supply is required to give energy in the form of ATP. Homeostatic mechanisms normally maintain blood glucose at a relatively constant level, through a negative-feedback loop. This promotes storage of glucose in the liver when there is excess in the blood and stimulates release of glucose by the liver into the blood as body cells use it up in respiration. This regulation is brought about by the actions of the pancreatic hormones, insulin and glucagon, as shown in the table.

Hormone	Produced in response to	Effect
Insulin	Increased blood glucose concentration after eating a meal	Liver cells respond by converting glucose into glycogen – blood glucose level decreases
Glucagon	Decreased blood glucose concentration between meals	Liver cells respond by converting glycogen into glucose – blood glucose level increases

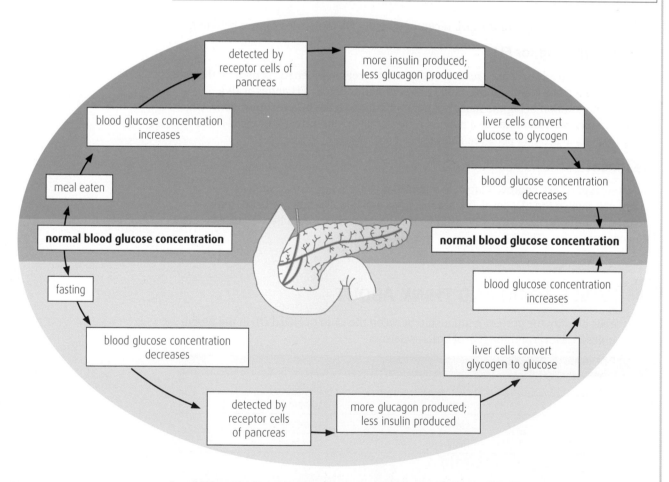

In addition, during periods of exercise and in the 'flight-or-fight' response, glucagon secretion is stimulated and insulin secretion is inhibited by the production of adrenalin (also called epinephrine) from the adrenal glands. This provides muscles with the glucose they require for increased activity.

In diabetes, this homeostatic mechanism breaks down – insulin production either stops or is decreased, causing the blood glucose level to remain at an elevated level (hyperglycemia). With elevated blood glucose level, more glucose enters endothelial cells than normal, causing the cells to malfunction and damaging blood vessels. Over time,

contd

atherosclerosis can develop in arteries (macrovascular disease). When small blood vessels are damaged, microvascular disease may occur, affecting the:

- retina – damaged blood vessels may haemorrhage causing blood to leak out and scar tissue to form, blocking vision and leading to blindness

- kidneys – leaky blood vessels allow protein to be excreted with the urine; prolonged damage causes blood vessels to collapse and kidney failure occurs

- peripheral nerves – damage to the nerves of the lower leg and foot (which detect temperature, pressure, and pain) causes numbness. This can allow damage to the area to go unnoticed, with skin ulcers and infection leading to possible gangrene and amputation.

A diabetes diagnosis is made using a **glucose tolerance test**. Here, the individual fasts for 8 hours before drinking between 250 and 300 ml of glucose solution. The blood glucose level is then measured over the next 2 hours. A graph showing the results of a glucose tolerance test is given below.

- In an individual with normal glucose metabolism, the glucose level rises after consumption and then falls rapidly as insulin is produced.

- People with **mild diabetes** have a relatively normal fasting blood glucose level. However after consumption, glucose level becomes much higher than in the normal individual and falls to the fasting level during the test period but at a low rate.

- In **severe diabetes**, fasting blood glucose level is much higher than normal. After consumption, it increases further and does not return to fasting level during the test.

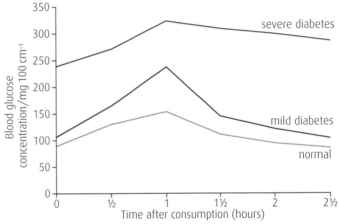

DON'T FORGET

An indicator of diabetes is the presence of glucose in the urine.

ONLINE

Learn more about diabetes by following the link at www.brightredbooks.net

Type of diabetes	Onset	Physiology and treatment
Type 1 diabetes	Usually in childhood	The pancreas does not produce any insulin. As a result, blood glucose level must be measured regularly and the correct amount of insulin injected.
Type 2 diabetes	Usually in adulthood; often in overweight individuals	The pancreas does produce insulin, but the cells have become resistant to it. Insulin resistance results from a decrease in the number of insulin receptors on the membranes of liver, muscle and fat cells. Binding between insulin and its receptor is important as it initiates movement of glucose transporters from an intracellular pool to the cell membrane, allowing glucose to pass into the cell. A decrease in the number of receptors, therefore, results in less glucose entering the cell.

THINGS TO DO AND THINK ABOUT

The diagram shows some of the stages in the control of blood glucose level.

1. increase in blood glucose level detected

↓

2. hormone X released

↓

3. glucose converted to substance Y

↓

4. blood glucose level decreases

1 In which organ would the increase in blood glucose level be detected?

2 Name hormone X and substance Y.

3 Where does stage 3 occur?

4 If hormone X was not present, the blood glucose level would drop very slowly. Why?

OBESITY

In obesity there is an excess of body fat compared to muscle (lean body tissue), which is indicated by a body mass index (BMI) of greater than 30 kg/m².

Obesity increases the chances of developing cardiovascular disease (CVD) and type 2 diabetes. It is caused mainly by:

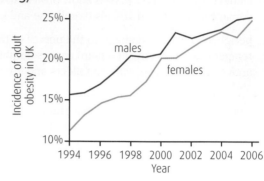

- a high calorie diet – fats contain twice the energy of carbohydrates and proteins per gram, and sugary foods are digested without any metabolic energy input

- lack of exercise – a sedentary lifestyle reduces energy expenditure.

The modern lifestyle in Western societies, including the UK, has resulted in an increase in the incidence of obesity. Treatment is primarily through dieting and exercise.

VIDEO LINK

Watch the clip at www.brightredbooks.net for more on underwater weighing.

A Bodpod being used to measure body volume

BODY COMPOSITION MEASUREMENT

The proportion of fat to lean tissue can be determined using a variety of methods. Some of these provide an accurate measurement but are difficult to carry out. Other methods are easy to perform but provide a less accurate estimate.

Densitometry

Fat has a lower density than muscle, so an individual carrying a greater percentage of body fat has a lower body density.

Body density is calculated by dividing body weight by volume and is measured in grams per cm³. An equation is then used to calculate percentage body fat.

While it is easy to measure body weight, measurement of volume is more difficult. This can be done by:

- placing the individual in a **BODPOD** and measuring air displacement

- **underwater weighing** – measuring the difference between the individual's weight on land and when submersed in water (1 g buoyancy = 1 cm³ volume).

Skin Fold Thickness

Skin fold calipers are used to measure skin fold thickness at specific sites on the body, including the front and back of the upper arm, below the scapula and above the hip. This method is quick and simple to carry out, and allows comparisons to be made over time. It can be difficult to take accurate measurements and the method does not take into account any natural differences in body fat distribution between people.

Bioelectrical Impedance Analysis (BIA)

Lean body tissue is a good electrical conductor (as it has a high water content), but fat (having a much lower water content) increases resistance or **impedance** to electrical current. By passing a small electrical current through the body and measuring the resistance, it is possible to determine the percentage body fat.

This method is not as accurate as other methods because the level of hydration and body temperature changes can also affect electrical conduction. In general, body fat is underestimated in individuals who are overweight, while overestimation in lean individuals is common.

contd

Waist:Hip Ratio

The waist:hip ratio is simply calculated by dividing the waist circumference by the hip circumference. Individuals who distribute body fat on their abdomen (and tend to store more fat around their internal organs) are said to be apple shaped and have a waist:hip ratio of greater than 0·8 (females) or 0·95 (males). If body fat is distributed more on the hips and thighs, the individual is said to be pear shaped and has a waist:hip ratio of less than 0·8 (females) or 0·95 (males).

Apple-shaped individuals are more at risk of developing heart disease and type 2 diabetes.

Body Mass Index

A body mass index (BMI) table is used to categorise an individual on a scale ranging from underweight to severely obese. BMI is calculated using the following equation:

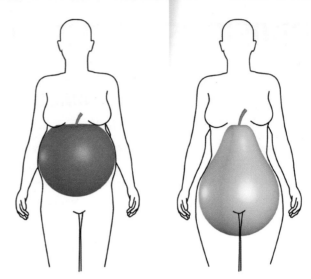

$$BMI = \frac{\text{weight (kg)}}{\text{height}^2 (m^2)}$$

A high waist:hip ratio gives an 'apple-shaped' body as there is extra fat around the waist. A 'pear-shaped' body has extra fat around hips and thighs and a low waist:hip ratio.

A BMI of 25–30 indicates a moderate risk of developing CVD, high blood pressure or type 2 diabetes. If the BMI is greater than 30, the risk increases further.

Although BMI is simple to calculate, it can give a poor indication of risk; individuals with a high body mass due to lean tissue (such as body builders) have a BMI which incorrectly indicates that they are overweight.

EFFECTS OF EXERCISE ON THE BODY COMPOSITION

Regular aerobic exercise (such as jogging, swimming or cycling) which increases the heart rate to between 55% and 70% of the maximum (220 – age) maintains a healthy lifestyle. This has important benefits through:

- increasing energy expenditure relative to energy intake, leading to loss of body fat

- increasing and maintaining lean muscle tissue and therefore increasing the muscle:fat ratio

- increasing the basal metabolic rate (BMR), as muscle has a higher metabolic rate than fat tissue.

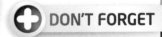 **DON'T FORGET**

A healthy diet is also important in reducing the risk of obesity and CVD.

 **DON'T FORGET**

Regular exercise can increase the HDL:LDL ratio, which also lowers the risk of CVD (see page 52).

THINGS TO DO AND THINK ABOUT

The graph below shows the incidence of obesity in children from 1971 to 2000. Describe two trends that can be observed from the graph.

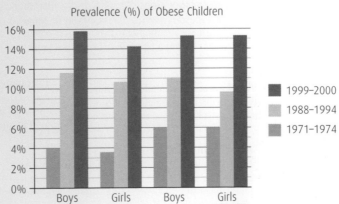

Prevalence (%) of Obese Children

Legend:
- 1999–2000
- 1988–1994
- 1971–1974

 ONLINE TEST

Head to www.brightredbooks.net to test your knowledge of obesity and body composition.

DIVISIONS OF THE NERVOUS SYSTEM

ONLINE

Learn more about the brain by following the link at www.brightredbooks.net

STRUCTURAL DIVISIONS OF THE NERVOUS SYSTEM

The nervous system consists of:

- central nervous system (CNS) – brain and spinal cord
- peripheral nervous system (PNS) – peripheral nerves.

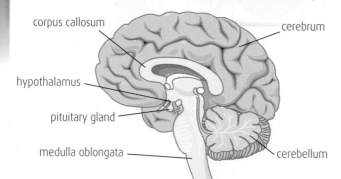

THE CENTRAL NERVOUS SYSTEM (CNS)

The central nervous system consists of the brain and spinal cord, and contains neurons (cell bodies, axons and dendrites) and their supporting cells (glial cells).

The Brain

The table shows the parts of the brain with which you should be familiar.

Part of brain	Function
Cerebrum	Involved with conscious activities, such as sensation. Also recalls memories and alters behaviour in light of experience. The left cerebral hemisphere receives sensory information from the right side of the body, including the visual field, and controls the right side of the body.
Cerebellum	Part of the central core of the brain, the cerebellum coordinates contraction of skeletal muscles and controls balance, posture and movement.
Medulla oblongata	Part of the central core, the medulla oblongata controls essential body processes, such as breathing, heart rate, arousal and sleep.
Hypothalamus	The hypothalamus is important in maintaining homeostasis and for regulating basic drives, such as sexual behaviour, drinking and eating.
Limbic system	The limbic system is involved in processing information for memories and influencing emotional and motivational states.

Cerebrum

The cerebrum consists of two cerebral hemispheres (left and right) that are connected by a bridge of nerve fibres called the **corpus callosum**. The corpus callosum is the only route of communication between the hemispheres. The surface of the cerebrum is folded, allowing more cell bodies to be present and maximising the number of connections between neurons.

Functional Areas of the Cerebrum

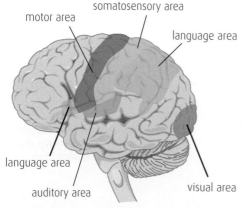

Somatosensory area: receives sensory information from the internal organs, skin and muscles. Parts of the body can be mapped out along the somatosensory area, with body parts that experience fine sensation (such as the lips and fingertips) having a larger area of the cerebrum.

Motor area Controls the contraction of specific muscles. Just like the somatosensory area, the parts of the body are organised along the motor area. Where muscles require fine control (such as muscles of the lips, tongue and fingers), they take up a greater proportion of the motor area. A **motor homunculus** can be drawn to show the relative size of the motor area that controls each part of the body.

Visual area: receives information from the eyes through the optic nerves. Interprets colour, shape and movement.

Association areas: concerned with imagination, personality, intelligence, problem solving and creativity.

Language areas: involved with controlling muscles required for speech (muscles of the lips, tongue and larynx) and memory of vocabulary.

Auditory area: receives information from the cochlea in the inner ear, through the auditory nerve. Interprets both pitch and rhythm.

FUNCTIONAL DIVISIONS OF THE PERIPHERAL NERVOUS SYSTEM

The peripheral nervous system can be divided into two functional parts.

1 Somatic Nervous System

The **somatic nervous system** is responsible for:

- voluntary control of skeletal muscles
- involuntary reflexes.

2 Autonomic Nervous System

The **autonomic nervous system** regulates the internal environment (for example, heart rate, body temperature and digestion). It has two divisions: **sympathetic** and **parasympathetic**. These perform opposing functions (and are, therefore, **antagonistic**), the sympathetic division preparing the body for action and the parasympathetic division returning the body to the resting state. For example, sympathetic nerves speed up the heart rate, while a parasympathetic nerve (the vagus nerve) slows down the heart rate (see page 47). Similarly, sympathetic nerves slow down the rate of peristalsis, and parasympathetic nerves speed it up again once the period of excitement is over.

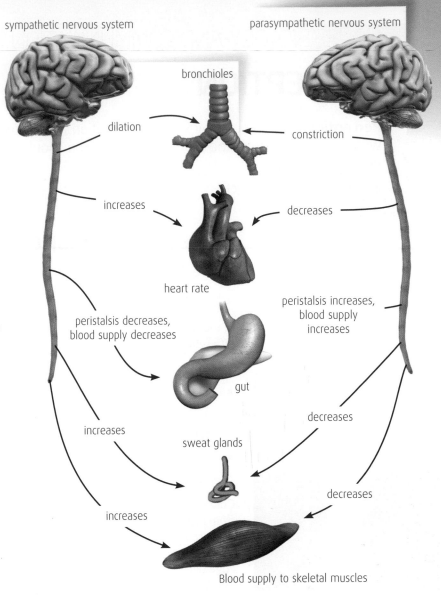

sympathetic nervous system   parasympathetic nervous system

bronchioles

dilation — constriction

increases — decreases

heart rate

peristalsis decreases, blood supply decreases — peristalsis increases, blood supply increases

gut

increases — decreases

sweat glands

increases — decreases

Blood supply to skeletal muscles

THINGS TO DO AND THINK ABOUT

In split-brain patients, the **corpus callosum** is cut, often to reduce epileptic seizures, preventing transfer of information between the right and left sides of the cerebrum. To understand the effect that this has on the patient, we must remember:

- Visual information in the left field of view is projected onto the right visual cortex, and information in the right field of view is projected onto the left visual cortex.
- The left motor cortex controls muscles on the right side of the body, and the right motor cortex controls muscles on the left side of the body.
- The speech area is usually found only on the left cerebral hemisphere.

In the experiment shown here, a picture of a ball is in the left field of view and a cube is in the right field of view. If asked what can be seen, the patient will say 'cube', as this is projected to the left hemisphere – communication between left visual area and speech area takes place. They cannot say 'ball' because communication between the right and left hemispheres has been cut.

If asked to use the left hand to pick up the ball or the cube from several hidden under a cloth, they would pick up the ball (communication between right visual and motor areas takes place) but wouldn't be able to tell you what they were holding.

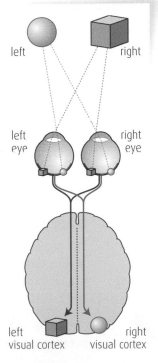

left — right

left eye — right eye

left visual cortex — right visual cortex

DON'T FORGET

The autonomic nervous system controls flight-and-fight responses and the peripheral nervous system controls rest and digest responses.

ONLINE TEST

Test yourself on divisions of the nervous system at www.brightredbooks.net

PERCEPTION

Perception is the process by which information from sensory receptors is analysed so that we can make sense of the world around us. You should be familiar with the three areas of visual perception:

- segregation of objects
- perception of distance
- recognition.

SEGREGATION OF OBJECTS

Figure–Ground Relationship

Our brain receives visual information about the whole environment and must make sense of this to differentiate between an object's shape (**the figure**) and the less distinct images in the background (**the ground**). To differentiate between figure and ground, we combine different visual cues:

- the figure resembles a meaningful object from memory
- the figure is seen as being in front of the ground
- the outline separating figure and ground is seen as belonging to the figure (**edge-assignment**)
- the figure is seen as less abstract in appearance than the ground.

Think about these cues as you look at the Edgar Rubin's vase shown here.

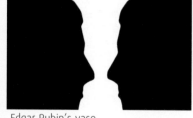
Edgar Rubin's vase

Concentrating on the white area, the image is seen as a vase from memory. It is perceived as being in front of the black background; and the outline around the white area is seen as the edge of the vase. Concentrating on the black areas, the edge between black and white now belongs to the black heads that can be seen in front of a white background.

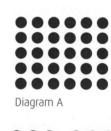
Diagram A

Coherent Patterns

When objects are close together, we perceive them as a group. In diagram A, the red dots are perceived as one group because they are evenly spaced. However, in diagram B, the red dots are perceived as belonging to two different groups because of the increased space between the third and fourth dot in each row.

Diagram B

PERCEPTION OF DISTANCE

Judgement of distance involves the use of **binocular disparity**, **visual cues**, and **perceptual constancy**.

Binocular Disparity

The position of objects in the image which is projected onto the retina is slightly different in the left and right eye, because each eye views the scene from a slightly different position. The images from each eye are merged by the visual cortex to produce a 3D image which gives an appreciation of distance. The amount of binocular disparity depends on the distance between the point of fixation and the second object, as shown in the diagram (note the distances between the two objects in each retinal image).

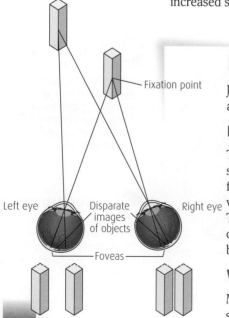

Fixation point

Left eye — Disparate images of objects — Right eye

Foveas

Visual Cues

Many visual cues assist in judging distance. You should be familiar with relative size, relative height in field of view, and superimposition.

contd

- When similar objects are viewed in a scene, their **relative size** depends on how far away they are. As the relative size of the buildings and telegraph poles decreases in the diagram, they are perceived as being further away.

- When we view a scene which lies below the horizon, objects that are positioned higher up and closer to the horizon are perceived as being further away than objects positioned lower down in the scene. This is called the **relative height in field of view**.

- When two objects overlap (**superimposition**), the object which is partially blocked from view is perceived as being further away.

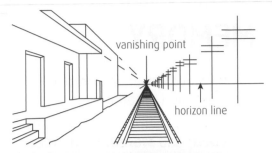

vanishing point

horizon line

Perceptual Constancy

The ability of humans to recognise objects as unchanged, even if the incoming sensory information has changed in terms of angle, size or colour, is called **perceptual constancy**. In the opening door example, we sense the shape of the door changing as it opens, but we perceive that the door remains the same rectangular shape.

In the photograph, the cars which are further away appear to sit higher up and closer to the horizon.

DON'T FORGET

Many visual cues assist in judging distance.

RECOGNITION

The ability to identify and give meaning to an object is called **recognition**. This involves many complex steps that take place in different parts of the cerebrum, including matching of the visual image to a mental representation stored in the **long-term memory**. The most important factor in this process is the **shape** of the object. In many cases, the information coming into the brain is incomplete and, through a process called **inference**, past experience is used to fill in any gaps. In addition, as a result of expectations, past experience and the object's context, the brain ignores some aspects of sensory information and focuses on others. This readiness to perceive the object as we expect it to be is called **perceptual set**.

VIDEO LINK

Watch the clip at www.brightredbooks.net for a great example of an optical illusion.

Example:

An example of perceptual set is shown opposite.

When you read the numbers from top to bottom, you expect the middle number to be 13. However, your expectation when reading the letters from left to right, is for the middle letter to be B.

ONLINE

Learn more about perception by working through the presentation at www.brightredbooks.net

THINGS TO DO AND THINK ABOUT

1 The ability to identify an object under various conditions is called:

   A superimposition   C recognition

   B perceptual constancy   D binocular disparity.

2 Which of the following is the most important factor in recognition of objects?

   A relative size   C coherent patterns

   B relative height in field   D shape

ONLINE TEST

Take the test on perception online at www.brightredbooks.net to test your knowledge of this area.

MEMORY

Memory involves all the processes that enable us to store, retain and retrieve information about sensations and emotions.

To form new memories, information coming into the brain is changed into a usable form (**encoded**) and **stored**. There must also be a method for **retrieval**, so that we can access the information at a later date.

DON'T FORGET

Shallow encoding results from repetition and elaborative encoding results when information is linked to previous memories.

ENCODING

During **encoding**, information that we see, hear, think and feel is changed into a form that can be stored as a memory. Encoding can be by:

- **acoustic code** – a sound image is created
- **visual code** – a visual image is created.

Storage

sensory memory	Information coming into the brain from the senses is stored very briefly (for between 0·5 and 3 seconds) in the sensory memory. Only a very limited amount of information from the sensory memory is retained and passed on to the next level of memory: short-term memory.

transfer ↓

short-term memory	This is the working memory and consists of information of which we are consciously aware. Short-term memory holds only a small amount of information (**7 ± 2 items**) for up to **30 seconds**. To prevent loss of information from the short-term memory, we can use **rehearsal** (repeating the information). Rehearsal helps with transfer of information into the next memory level: long-term memory. Information that is organised into categories during encoding has a greater chance of being transferred to the long-term memory. Information that is not transferred to the long-term memory will be lost (**displacement or decay**).

item displaced ← rehearsal

transfer ↓

long-term memory	The long-term memory has an unlimited capacity and stores information outside our conscious awareness for a long period of time. Information is grouped by type of memory (facts or skills for example).

Short-Term Memory

The short-term memory span is 7 ± 2 items. However, we can boost memory by putting related information into groups (**chunking**), forming larger meaningful items that can be stored. For example, the letters and numbers ITV1BBCCH4CH5ITV2MTV form 20 items if taken individually but if we chunk them to get ITV1-BBC-CH4-CH5-ITV2-MTV they form six meaningful items that can be memorised.

Rehearsal is used to aid transfer of information from short-term to long-term memory. Its role can be demonstrated using the **serial position effect**. An experiment to look at this effect is carried out as follows:

1 Subjects are asked to put pens and pencils down before the items are revealed (to prevent cheating).

2 Twenty items are shown, one at a time, each for 5 seconds (limiting rehearsal time).

3 After all 20 items have been shown, subjects write down as many as they can recall.

4 The experiment is repeated with a fresh set of items (to increase reliability).

The graph on the next page shows a typical set of results for this experiment.

contd

Items at the beginning of the sequence are remembered well because there has been time for rehearsal and transfer to the long-term memory. Items at the end of the sequence have not been displaced from the short-term memory and are, therefore, also recalled well. However, items in the middle of the sequence are poorly recalled as they have been displaced from the short-term memory.

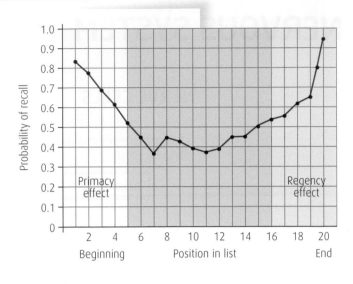

Long-Term Memory

Transfer of information into the long-term memory can be made more successful by **elaborating meaning**. This works by devising a little story about the item to be remembered – making the information stand out.

Retrieving information from long-term memory is made easier if you are in the same setting or context as you were when the information was encoded. Here, particular sights, sounds, smells or emotions act as **contextual cues**, triggering retrieval of a memory. For example, visiting your old home or school can evoke memories of your childhood, or the smell of a perfume as you enter a room can evoke memories of a particular person or an event.

Location of Memory in the Brain

Parts of the brain associated with storage of memories include areas of the cerebral cortex and the **limbic system**. The limbic system lies deep in the brain, extending into the temporal lobe of the cerebrum.

The site of long-term memory storage depends on the type of memory.

Type of memory	Storage location	Example
Procedural	motor cortex	motor and cognitive skills (e.g. how to ride a bike)
Emotional	limbic system	positive and negative emotional experiences
Spatial	hippocampus of the limbic system	information relating to the surrounding environment and spacial orientation of objects within it
Episodic	area of the cerebral cortex in which the sensory information was encoded	memories of personal experiences and events
Semantic	area of the cortex where the information was encoded	general knowledge (facts), and abstract ideas and concepts

 DON'T FORGET

Contextual cues are built up during encoding; elaboration adds to the number of contextual cues for each piece of information. The more contextual cues that are present, the easier the recall.

 THINGS TO DO AND THINK ABOUT

Short-Term Memory Span
You should be familiar with the method used to determine the capacity of the short-term memory.
1. Subjects are asked to put pens and pencils down before starting the experiment (to prevent cheating).
2. A series of three letters or numbers is read out one at a time, in a monotone voice and at regular speed.
3. Subjects pick up pens and pencils, and write down the series.
4. The procedure is repeated with the number of items in the series increasing each time.
5. A second set of numbers is then used to increase reliability.
6. The maximum number of items that can be recalled correctly by each subject is taken as their memory span, with class results being pooled to determine the maximum and minimum memory span for the class.

A typical set of results would show the memory span to range from 5 to 9 items, ie 7 + 2.

 ONLINE

Learn more about memory by following the links at www.brightredbooks.net

 ONLINE TEST

Test your knowledge of memory at www.brightredbooks.net

NERVOUS SYSTEM

NERVE CELLS

The functional cell of the nervous system is the nerve cell or **neuron**. There are three types of neuron: **sensory**, **motor** and **inter neuron**. Each neuron consists of the following parts:

> **Dendrite** Dendrites receive nerve impulses and send them towards the cell body.

⬇

> **Cell body** The cell body contains the nucleus.

⬇

> **Axon** The axon carries the nerve impulse away from the cell body.

DON'T FORGET ➕

The cell body of a sensory neuron sticks up from the rest of the nerve like a little lollipop.

Sensory neurons

Sensory neurons pass information from **sense receptors** to neurons in the central nervous system

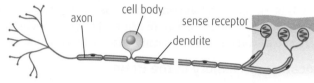

(CNS). A single dendrite receives information from sense receptors and transmits a nerve impulse towards the cell body. From the cell body, a single axon carries the nerve impulse into the spinal cord where the impulse is transmitted to the dendrites of an **inter neuron**.

Inter Neurons

Inter neurons lie completely within the CNS. They vary in shape but, in general, have several dendrites and an axon extending from the cell body, allowing messages to be passed on to a large number of neurons.

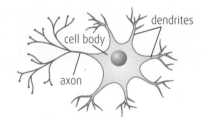

Motor Neurons

Motor neurons transmit nerve impulses from the CNS to an effector organ (muscle or

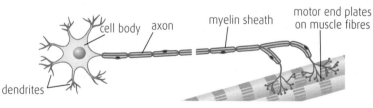

gland). The cell body and several short dendrites lie embedded in the CNS. One axon extends from the cell body, passing out of the CNS to reach the effector organ.

DON'T FORGET ➕

Receptors on the post-synaptic membrane determine whether the signal is excitatory or inhibitory.

THE SYNAPSE

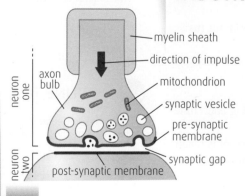

A synapse is the junction between an axon of one cell and a dendrite of another, and allows the transmission of nerve impulses between neurons.

The nerve impulse passes down the axon to reach the **axon bulb**, where it stimulates **synaptic vesicles** to move towards and fuse with the **pre-synaptic membrane**, releasing neurotransmitter chemicals by exocytosis. Neurotransmitters diffuse across the **synaptic gap** to reach the **post-synaptic membrane**, where they fuse with receptors. If a large enough number of receptors are activated, the post-synaptic membrane reaches its **threshold** and the nerve impulse is transmitted onwards.

Neurotransmitters must be rapidly removed from the post-synaptic membrane. Otherwise, it would be impossible to control the frequency of

contd

nerve impulses, and you would not be able to distinguish between stronger and weaker stimuli (such as bright and dim light).

Neurotransmitter Substances

You should be familiar with two neurotransmitter substances:

Acetylcholine is removed by enzymatic degradation. The enzyme acetylcholinesterase breaks down the neurotransmitter and the inactive products are reabsorbed by the pre-synaptic cell. They are recycled and used to make more acetylcholine.

Noradrenaline leaves the post-synaptic membrane and is reabsorbed intact by the pre-synaptic cell. Within the pre-synaptic cell, noradrenaline is enclosed within vesicles for reuse.

DON'T FORGET

If not enough neurotransmitter is released to reach the threshold, the nerve impulse will not be transmitted. However, a series of weak stimuli can together allow that threshold to be reached in a process called **stimulation**.

MYELINATION

The speed of conduction of a nerve impulse is increased by the neuron's **myelin sheath**. **Myelin** is formed by **glial cells** which wrap themselves round and round axons, building up layers of cell membrane.

Myelination is incomplete at birth but continues as the child grows, being completed in the upper limbs before the lower limbs. As a result, babies cannot coordinate their movements at birth and gain control of their arms before their legs.

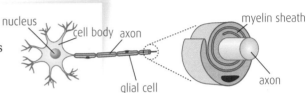

NEURAL PATHWAYS

The route that a nerve impulse follows through the nervous system is called a **neural pathway**. New neural pathways are formed to allow new responses to stimuli, suppress reflexes or to bypass tissue damage. This ability to remodel is called **plasticity**. You should be familiar with three types of neural pathway: converging, diverging and reverberating neural pathways.

ONLINE

Follow the link to the animation of the synapse at www.brightredbooks.net to learn more.

Converging Neural Pathways

In a **converging neural pathway**, impulses from different sources are directed to one neuron. This allows weak stimuli to be amplified, as in the visual pathway. Within the retina, the rods that detect low light release only a small amount of neurotransmitter into the synaptic gap. However, a cumulative effect results through the synapse of several rods with the same post-synaptic neuron, resulting in the transmission of a nerve impulse (**summation**).

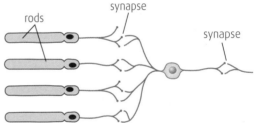

Diverging Neural Pathways

In a diverging pathway, a single pre-synaptic neuron forms synapses with several post-synaptic neurons, allowing the pathway to reach several destinations.

Diverging neural pathways allow fine motor control, where several skeletal muscles work together to produce precise movements, such as in the fingers or eyes.

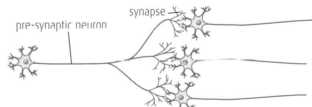

Reverberating Neural Pathways

In a reverberating pathway, neurons link back to form synapses with neurons earlier in the chain, so that the nerve impulse passes repeatedly through the circuit. An example is the neural pathway that controls breathing.

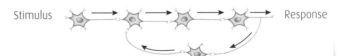

THINGS TO DO AND THINK ABOUT

Glial cells do more than produce the myelin sheath. Describe the other roles that they perform.

ONLINE TEST

Take the test on the nervous system online at www.brightredbooks.net

NEUROTRANSMITTERS AND RECREATIONAL DRUGS

Neurotransmitters are chemicals that transmit nerve impulses across synapses from one neuron to the next (see previous page). Some neurotransmitters are involved in controlling mood. These include:

endorphins – involved in pain relief and give a feeling of euphoria (similar to the effect of **opiates**)

dopamine – involved with feeling pleasure.

ENDORPHINS

There are about 20 different **endorphins**. They function by attaching to receptors on neurons, **blocking the transmission of pain signals**. The increased production of endorphins in response to very severe pain enables continued body function while the individual finds help, increasing the chances of survival. Endorphins also have a role in the regulation of appetite and cause the release of sex hormones.

Endorphins are produced under other circumstances:

- During extended periods of physical exercise, endorphin production increases, producing feelings of euphoria. Their action inhibits the production of **GABA** (an inhibitory neurotransmitter), unblocking the **reward pathway**. Dopamine is produced as a result of increased endorphins in the reward pathway.

- physical and emotional stress

- during eating of some foods, such as dark chocolate.

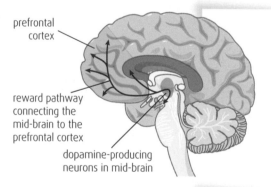

prefrontal cortex

reward pathway connecting the mid-brain to the prefrontal cortex

dopamine-producing neurons in mid-brain

DOPAMINE

When humans participate in activities that are rewarding (involving natural rewards such as food or artificial ones such as drugs), the pleasurable feelings act as positive reinforcement, making it more likely that the behaviour will be repeated. The **reward pathway** which produces these feelings of pleasure connects dopamine-producing neurons in the **mid-brain** with neurons in the **prefrontal areas** and **base of the cerebral cortex**. Dopamine also has a role in controlling motor function through a separate pathway.

TREATMENT OF NEUROTRANSMITTER DISORDERS

When there is either an overproduction or underproduction of neurotransmitter, or if there is an imbalance in neurotransmitter production, a neurotransmitter disorder can result. These include:

- Alzheimer's disease
- Parkinson's disease
- schizophrenia
- anxiety disorders
- depression

To treat these disorders, drugs that are similar to neurotransmitters are prescribed. These act as **agonists**, **antagonists** and **enzyme inhibitors**.

Agonists

Agonists are drugs that copy the function of a neurotransmitter. They bind to the same receptor as the neurotransmitter, bringing about a normal response. Dopamine agonists are used in the treatment of Parkinson's disease.

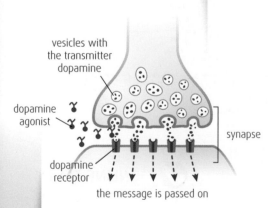

vesicles with the transmitter dopamine

dopamine agonist

dopamine receptor

synapse

the message is passed on

Antagonists

Antagonists are drugs that prevent the neurotransmitter binding to its receptor on the post-synaptic membrane, by competing with the neurotransmitter for the receptor binding site or by causing a change in shape of the receptor, so that the neurotransmitter cannot bind. Dopamine antagonists are used to block dopamine receptors when treating schizophrenia, which is caused – in part – by the overstimulation of dopamine receptors in the reward pathway.

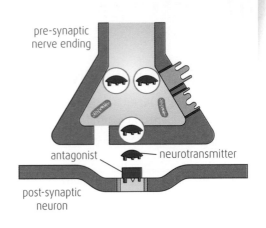

pre-synaptic nerve ending

antagonist — neurotransmitter

post-synaptic neuron

Enzyme Inhibitors

Neurotransmitters are normally removed from the post-synaptic membrane by enzyme action or by re-uptake into the pre-synaptic neuron. Enzyme inhibitors work by preventing this. For example, in the treatment of Alzheimer's disease, an enzyme inhibitor slows down acetylcholinesterase, the enzyme which breaks down acetylcholine, maintaining its concentration.

RECREATIONAL DRUGS AND ADDICTION

Recreational drugs often mimic the action of neurotransmitters in the reward pathway, altering brain neurochemistry and changing mood, cognition, perception and behaviour. The heightened release of dopamine in the reward pathway provides the motivation to repeat the act, often leading to addiction. As with drugs used to treat neurotransmitter disorders, recreational drugs can act as:

- agonists – ethanol binds to GABA receptors on the post-synaptic membrane causing widespread changes to behaviour; nicotine is a dopamine agonist, binding to dopamine receptors to produce feelings of a 'dopamine high'

- antagonists – ethanol blocks the glutamate receptor NMDA, which is involved in memory

- inhibitors – cocaine blocks the mechanism which removes excess dopamine and serotonin from the synapse by binding to the molecules that transport them across the synaptic cleft. They cannot, therefore, be reabsorbed into the pre-synaptic neurons.

transmitting neuron

dopamine transporter blocked by cocaine

cocaine

receiving neuron

dopamine receptor

intensity of effect

dopamine

Addiction

When drugs that act as **antagonists** are abused over time, the body reacts to the reduced stimulation of receptors by increasing the number of receptors and their sensitivity to the drug. This is called **sensitisation** and leads to **addiction**.

Long-term use of drugs that act as **agonists** causes a decrease in the number and sensitivity of receptors for the drug (**desensitisation**). Larger and larger doses of the drug are needed for an equivalent effect (**drug tolerance**).

 THINGS TO DO AND THINK ABOUT

Research online the mode of action of the recreational drug Ecstasy.

 ONLINE

Learn more about how Alzheimer's disease affects the brain at www. brightredbooks.net

DON'T FORGET

Sensitisation leads to addiction. Desensitisation leads to drug tolerance.

ONLINE TEST

Revise your knowledge of neurotransmitters and recreational drugs at www. brightredbooks.net

 ONLINE

Follow the link at www. brightredbooks.net for more on this topic.

BEHAVIOUR

EFFECT OF EXPERIENCE

Learning happens when human behaviour is modified in the light of experience. Experience is gained in several ways.

Practice

The repeated use (practice) of **motor skills** causes an increased number of synapses to form in the brain, producing a **motor pathway**. Skills, such as writing or driving a car, develop through the formation of this **motor memory**.

Imitation

In many situations, the preferred method of learning is through imitation. Here, a behaviour is observed and then copied. For example, children may learn to cross the road by copying adults. Imitation is often used during **training**.

Trial and Error

When an individual is **rewarded** for performing a particular behaviour (for example, a parent smiles and interacts with an infant in response to a new word), the behaviour is **reinforced** and is likely to be repeated. When the behaviour is not rewarded, it may become **extinct**.

Individuals can be trained to display positive behaviour through **shaping.** Here, each time performance moves closer to the desired behaviour, a reward is given, until – eventually – the desired behaviour is displayed. This technique can be used, for example, when children are toilet-training.

Generalisation and Discrimination

Sometimes an individual responds in a similar manner to different but related stimuli (for example, a fear of *all* spiders). This indiscriminate response is called **generalisation**. However, when the individual is able to give a different reaction to related stimuli (they are scared of big spiders, but not small ones), the response is termed **discrimination**.

GROUP BEHAVIOUR AND SOCIAL INFLUENCE

An individual's emotions and behaviour are influenced by the presence of others.

Identification

Individuals alter their behaviour to be more like someone they admire in a process called **identification**. This role model may be a family member, friend or colleague, a celebrity or sports star. Identification is used by advertisers who employ celebrities to sell their products.

Deindividuation

Deindividuation involves the loss of personal identity in a group situation. Having become anonymous within the group, individuals lose their sense of judgement and behave in a less appropriate manner (as seen in the case of anti-social behaviour displayed by football hooligans).

Internalisation

Internalisation is a change in an individual's beliefs due to **persuasion**. Governments use persuasion in their health campaigns.

Social Facilitation

Individuals often perform better in tasks when they are in the company of others (either other competitors or an audience). For example, individuals often run faster when competing against others in a race than when they run alone.

THINGS TO DO AND THINK ABOUT

A finger maze can be used to construct a learning curve (shown below), either by measuring the time taken to complete the task, or by counting the number of errors made when completing the task.

The graph shows that, as the number of trials increases (and the subject practises the task), performance improves until the task cannot be carried out faster.

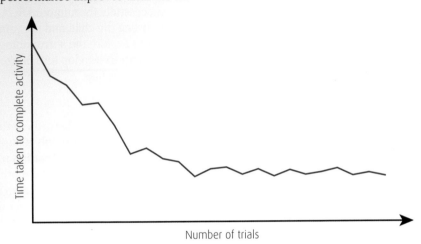

1 Explain why each of the following factors should be considered in the design of a finger-maze experiment:
   a width of the maze
   b which finger is used in the experiment
   c whether the subject wears a blindfold
   d number of trials per subject
   e number of subjects used in the experiment.

2 An experiment was carried out to find out how many sit-ups a pupil could do in one minute. It was found that pupils can do more sit-ups when tested in a group than when tested alone. What term is used to describe this effect?

3 A child who was bitten by a labrador now responds with fear to all dogs. What term is used to describe this effect?

4 When a person changes the washing powder they use because their sporting idol advertises the product, this is an example of which type of influence?
   A social facilitation
   B internalisation
   C deindividuation
   D identification.

 VIDEO LINK

For more about group behaviour, watch the clip at www.brightredbooks.net

 DON'T FORGET

It is important that sufficient trials are included in the experiment to allow the learner to improve to their maximum ability. This would normally involve using a minimum of 10 trials.

 ONLINE TEST

Head to www.brightredbooks.net and test your knowledge of behaviour.

COMMUNICATION AND SOCIAL BEHAVIOUR

INFANT ATTACHMENT

In humans, there is a long period of dependency, when children develop social and communication skills that will allow them to function successfully throughout life. A strong emotional bond (**infant attachment**) develops between the child and the primary carers, providing the child with a secure base from which to explore the surroundings. Children who form secure attachments in infancy are thought to develop into more resilient and trusting adults who are able to form stable relationships.

The 'Strange Situation'

The 'strange situation' is a laboratory procedure developed by Mary Ainsworth. In this test, an infant between the ages 12 to 18 months is placed in a room with only its mother and is allowed to explore. A stranger then enters the room and has a conversation with the mother. The stranger then moves their attention to the infant, allowing the mother to leave the room unnoticed (a separation episode). The mother re-enters the room and comforts the child (a reunion episode), before leaving again. This time the stranger also leaves and the infant is alone (a second separation episode). The stranger then re-enters the room and interacts with the infant. Finally, the mother returns and comforts the infant (a second reunion episode). The infant's behaviour is scored based on how much they explore, their reaction on separation from the mother and on being reunited with her, and also on the level of anxiety when alone with the stranger.

A securely attached infant would:
- explore the room and interact happily with the stranger when the mother is present
- become distressed when the mother is not present
- once reunited, become content and resume exploring the room.

Insecure attachment falls into three categories:
- avoidant insecure attachment – the infant does not explore much and does not treat the mother differently to the stranger; when the mother leaves it has little effect on the infant
- resistant insecure attachment ('clingy' child) – the infant doesn't explore much, becomes distressed when the mother leaves and is wary of the stranger; on reunion with the mother, the infant is 'distant' and may become angry at having been separated
- disorganised insecure attachment – the infant displays aspects of both avoidant and resistant insecure attachment and seems anxious and confused.

Parenting Styles

The style of parenting that an infant is exposed to is important in determining their attachment status.

Permissive control: The parent provides little control over the child and seems indulgent, giving the child an extreme amount of freedom. The child will often insist on the encouragement of the parent during exploration and play. In addition, the child will show stranger anxiety and becomes easily distressed when the parent leaves the room.

Authoritative control: Here the carer blends a high degree of control and warmth, making choices and giving direction, but taking into account the child's own preferences. For example, the carer may choose to take the child to a park but the child chooses the piece of equipment they play on. This produces a confident child who will happily explore the world around them and has a secure attachment.

COMMUNICATION

The human species displays complex behaviour due to our highly developed language skills. Language is backed up through the use of **non-verbal communication**; a series of actions that may obey **social** and **cultural rules**.

DON'T FORGET

In the 'strange situation' experiment, the infant must be able to identify the stranger as 'foreign'.

Language

Language involves the use of symbols, such as words (spoken or written), to represent information. Through this, we organise thoughts and communicate them to others, speeding up our rate of learning and intellectual development. The ability to use language and communicate verbally has allowed humans to transmit knowledge and develop our culture.

Non-Verbal Communication

When we interact with each other, wordless signals are passed between us. These indicate attitudes and emotions, and add to any verbal message that they accompany. **Non-verbal communication** is, therefore, important in forming and maintaining all successful relationships, including the relationship between an infant and its carers – which must be based initially on non-verbal communication as the infant's language skills have yet to develop (think of the effect that an infant's smile has on the mother and father).

Some forms of non-verbal communication:

Non-verbal communication can either **reinforce** or **contradict** spoken language. If the non-verbal message is in agreement with the verbal message, it helps to make the message clear. For example, banging your fists on a table when shouting at someone reinforces anger; smiling while praising someone reinforces the praise.

Personal space: The physical distance that is maintained between two people indicates the level of intimacy between them. Close friends and family may be allowed to invade our personal space, but we generally keep others at a distance.

Facial Expression: We use a vast range of facial expressions to convey emotions such as:
- frowning – disapproval
- smiling – interest or attraction
- winking – humour.

Tone, volume and pitch of voice: Non-verbal aspects of speech add to the emotions conveyed by the verbal message. For example:
- A low volume can convey nervousness.
- A high pitch can signal excitement.

Gestures: Some gestures can substitute for verbal language (a nod can replace a verbal 'yes'; waving signals 'hello' in many cultures). Other gestures indicate emotions (a clenched fist can indicate anger).

Eye contact: We use eye contact (both frequency and length of gaze) to signal our emotions, to define our status and to regulate interactions between us. For example:
- If you are attracted to an individual, you will repeatedly try to catch their gaze.
- We can indicate anger by continuously staring at someone.
- Subservient individuals keep their eyes down, avoiding eye contact with someone they consider to be superior.

Body posture: We convey our attitude through the posture that we maintain. For example:
- Students convey that they are paying attention by sitting upright and not slouching
- We convey interest in another individual by angling our bodies towards them.
- We show that we are relaxed by sitting back with arms unfolded and legs extended.
- Friendship can be conveyed by mimicking an individual's body posture.

However, if the non-verbal and verbal messages are not in agreement, the message becomes confused. For example, by fidgeting and avoiding eye contact while telling the truth, our body language implies that we are lying.

We sometimes deliberately use contradictory non-verbal and verbal messages to either unnerve individuals or in humour when teasing someone.

THINGS TO DO AND THINK ABOUT

1 Explain why there is a long period of dependency in humans.

2 Non-verbal communication is important in infancy. (i) Explain why and (ii) give two examples of non-verbal communication that takes place between a mother and her infant.

ONLINE

Follow the link at www. brightredbooks.net for a case study on this topic.

DON'T FORGET

Matching verbal and non-verbal messages lead to stronger, easily understood communication.

ONLINE TEST

Test yourself on communication and social behaviour at www. brightredbooks.net

IMMUNOLOGY

Humans often live in dense populations and, so, are at risk from transmitted diseases. **Immunology** is the study of the *immune system,* both when healthy and during disease. The **immune system** is a complex system of biological structures and *processes* which detect and destroy a wide variety of pathogens (bacteria and viruses) and so protects against *disease*. The human body can also protect itself against some toxins and cancer cells via the immune system. **Immunity** is the ability of the body to resist infection by a pathogen or to destroy it if it invades the body.

NON-SPECIFIC DEFENCES

Physical and Chemical Defences

The body has natural barriers to prevent entry by pathogens. **Epithelial cells** form the outer covering of skin and line the cavities of the body. They form a physical barrier and have specialised cells which produce **secretions** such as mucus, wax or oil to trap dirt and pathogens. Various glands in the skin and the tear glands produce chemicals with antimicrobial properties. Tears, saliva and sweat contain a powerful enzyme, lysozyme, which breaks down the cell walls of airborne pathogens. The mucus secreted by the epithelial cells of the respiratory tract and the upper gastrointestinal tract is sticky and so traps foreign particles, which are then swept upwards (as in the respiratory tract) or destroyed by **phagocytes**. Hairs in the nose trap foreign particles, and the cough and sneeze reflexes remove foreign bodies from the throat and nose. Stomach acid destroys many pathogens taken in with food.

First lines of defence

saliva — antibacterial enzymes

tears — antibacterial enzymes

skin — prevents entry

mucus — linings trap dirt and microbes

stomach acid — low pH kills harmful microbes

"good" gut bacteria — out compete bad

Inflammatory Response

The inflammatory response is the body's reaction to injury or infection by pathogens and is localised to the site of injury or infection. Special white blood cells, called **mast cells**, present in the connective tissues of the body and in organs, play a key role. If tissue damage occurs, mast cells are activated and release the chemical **histamine**. Histamine increases the permeability of capillaries and causes blood vessels to dilate (**vasodilation**), increasing blood flow to the site of infection.

The increase in the flow of blood to the damaged tissue, as well as the secretion of **cytokines** by white blood cells in the damaged area, results in an increase in the number of phagocytes, antimicrobial proteins and clotting factors.

- Cytokines stimulate the movement of phagocytes to the site of infection.
- The phagocytes engulf and digest the pathogens by **phagocytosis**.
- The antimicrobial proteins amplify the **immune response**.
- The increase in blood-clotting chemicals promotes the **coagulation** of blood at the damaged tissue, halting blood loss and preventing further infection.

DON'T FORGET

The physical barriers and their secretions are the human body's first line of defence.

DON'T FORGET

The trachea and bronchi of the respiratory tract are lined with cilia which beat rhythmically to move mucus containing trapped microbes upwards to the mouth, where it can be swallowed.

DON'T FORGET

The increased blood flow that occurs as part of the inflammatory response results in heat and redness of the damaged tissue.

DON'T FORGET

Cytokines are small proteins which are released by certain cells and affect the behaviour of others via receptors.

contd

The Inflammatory Response

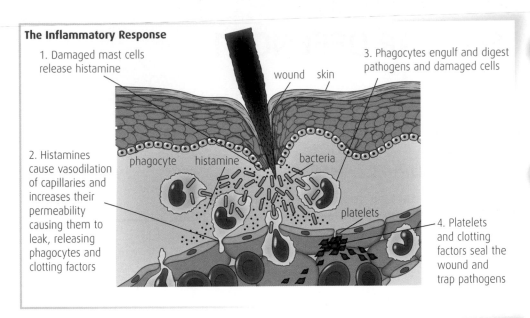

1. Damaged mast cells release histamine

2. Histamines cause vasodilation of capillaries and increases their permeability causing them to leak, releasing phagocytes and clotting factors

3. Phagocytes engulf and digest pathogens and damaged cells

4. Platelets and clotting factors seal the wound and trap pathogens

wound   skin

phagocyte   histamine   bacteria   platelets

WHITE BLOOD CELLS

There are several types of white blood cell which are part of the immune system and specialised to protect the body from infection by pathogens.

1 Phagocytes

The surface of every cell is covered with unique **antigen** molecules, usually proteins. Phagocytes recognise these antigens on the surface of pathogenic cells and destroy them by phagocytosis.

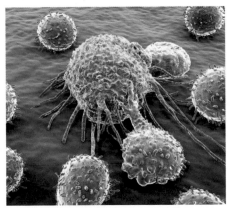

A phagocyte carrying out phagocytosis

cell processes   bacterium   vacuole

phagocyte   lysosome

Cell processes surround the bacterium, enclosing it in a vacuole.

The vacuole moves into the cell.

A lysosome fuses with the vacuole, releasing digestive enzymes.

Enzymes break down the bacterium. Products may be reused by the cell.

2 Natural Killer (NK) Cells

NK cells kill cancer cells and cells infected with viruses. They are non-specific and attack cells by releasing toxic proteins which **perforate** the membrane of the target cell and by releasing **signal molecules** which enter the cells via the perforations. These signal molecules stimulate the cells to produce enzymes which cause their own destruction: **apoptosis** or programmed cell death.

Both phagocytes and NK cells release cytokines which stimulate the specific immune response.

Natural killer cells attack a cancer cell

 THINGS TO DO AND THINK ABOUT

1 Describe two ways in which epithelial cells provide defence against pathogens.

2 Explain the role of the following in the inflammatory response:
   a mast cells   **b** histamine   **c** phagocytes   **d** platelets.

3 What is apoptosis? Explain the role of natural killer cells in apoptosis.

SPECIFIC CELLULAR DEFENCE 1

As well as non-specific defence mechanisms against disease, humans have an additional and sophisticated mechanism which is capable of recognising and destroying **specific** substances.

IMMUNE SURVEILLANCE

Immune surveillance is a theory which suggests that a range of white blood cells constantly circulate within the body, monitoring for damage to tissues or invasion by pathogens. They may also detect cells which have become cancerous. In response to damage or infection, some white blood cells produce **cytokines** which increase the blood flow to the site of damage or infection, resulting in the accumulation of white blood cells, including phagocytes (involved in non-specific defence) and **lymphocytes** (specific defence). Lymphocytes are a group of specialised white blood cells including natural killer cells, T cells and B cells.

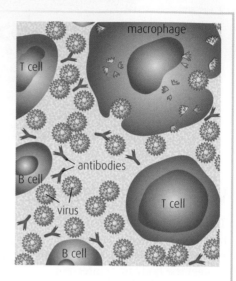

CLONAL SELECTION THEORY

The body has a huge number of different lymphocytes, each with a different membrane receptor that is specific for one particular antigen.

Stages in Clonal Selection

Stage 1: An antigen binds to its specific receptor on a lymphocyte.

Stage 2: The specific lymphocyte undergoes repeated division, resulting in the formation of a clone of identical lymphocytes.

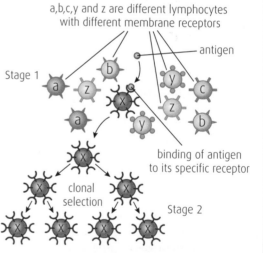

The clonal selection theory of antibody production

CELL-MEDIATED IMMUNITY

T Lymphocytes

T lymphocytes are produced in the bone marrow and then migrate through the bloodstream to the thymus gland in the upper chest area of the body, where they mature. T Lymphocytes play a major role in **cell-mediated immunity**. In the cell-mediated response, there is a direct interaction between the T lymphocytes and invading pathogen. T lymphocytes can be distinguished from other lymphocytes by their cell-surface receptors (T-cell receptors or TCR) which allow them to detect specific foreign substances (antigens) that enter the body. These specific surface proteins allow the T lymphocytes to distinguish between the body's own cells (self) and foreign cells (non-self).

contd

1 Cytotoxic T cells (also called killer T cells) destroy cells which the immune system regards as foreign (infected cells and tumour cells). They recognise antigens on these foreign cells, bind to the antigen and destroy the cells by **apoptosis**. Once cytotoxic T cells have completed their job, the majority undergo apoptosis themselves, but a few become **memory cells** and are involved in **immunological memory**.

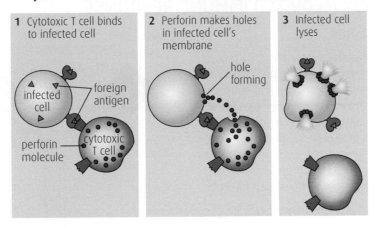

2 Helper T cells do not destroy infected cells, but secrete cytokines which activate B lymphocytes (see p76) and phagocytes.

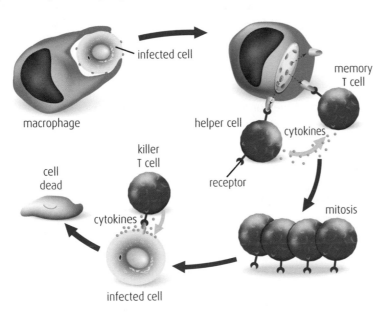

Antigen-Presenting Cells

When a tissue has been infected by a particular pathogen, phagocytes are attracted to the pathogens by chemicals and engulf them by the process of phagocytosis. Some phagocytes may then move part of the pathogen back to their own cell surface. These phagocytes are called **antigen-presenting cells** and can activate the production of a clone of T lymphocytes that move to the site of infection under the direction of cytokines.

THINGS TO DO AND THINK ABOUT

1 Describe what is meant by immune surveillance.

2 Write a few sentences to explain the clonal selection theory of antibody production.

3 Draw a table to show the functions of the three different types of T lymphocytes.

VIDEO LINK

Watch the video which describes the roles of different T cells at www.brightredbooks.net

VIDEO LINK

Watch the video clip about fighting infection by clonal selection at www.brightredbooks.net

DON'T FORGET

The immune response is specific and is the body's third line of defence.

ONLINE

Read about the action of T and B lymphocytes at www.brightredbooks.net

ONLINE TEST

Take the test on specific cellular defence at www.brightredbooks.net

SPECIFIC CELLULAR DEFENCE 2

DON'T FORGET

Helper T cells do not destroy infected cells but secrete cytokines which activate B lymphocytes and phagocytes.

DON'T FORGET

The clonal selection theory explains how the body can produce such a wide range of different antibodies.

ONLINE

Read more about the immune response at www. brightredbooks.net

ANTIBODY-MEDIATED IMMUNE RESPONSE

B Lymphocytes

B lymphocytes are involved in the humoral or **antibody-mediated immune** response. B lymphocytes are produced and mature in the bone marrow. B lymphocytes which are activated by antigen-presenting cells and T lymphocytes (see page 75) produce a **clone** of B lymphocytes which secrete **antibodies** into the lymph and blood systems, where they make their way to the site of infection.

Each clone of B lymphocytes (see clonal selection theory, page 74) produces a **specific** antibody which recognises a specific antigen surface molecule on a pathogen or toxin. For example, one type of B cell will make antibodies against the virus that causes the common cold, while a different type will produce antibodies that attack bacteria which cause pneumonia. An antibody matches an antigen much like a key matches a lock. The antibody binds to the antigen to form an antigen–antibody complex which:

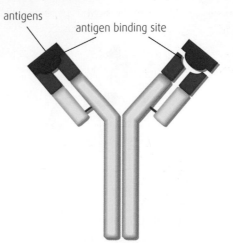

An antibody is a y-shaped protein. Each tip of the y has an antigen-binding site specific to a particular antigen.

- can inactivate the pathogen or toxin
- may make it more susceptible to phagocytosis
- may stimulate a response by T cells, resulting in cell lysis (cell destruction).

IMMUNOLOGICAL MEMORY

Some T and B lymphocytes produced in response to antigens by **clonal selection** survive long term as **memory cells**. Immunological memory is the ability of the immune system to respond more quickly and more effectively to a subsequent infection by the same antigen. A secondary exposure to the same antigen quickly brings about production of a new clone of lymphocytes, giving a more rapid and greater immunological response.

Immunological memory is specific for a particular antigen and is long-lived.

Memory B cells exist long after the infection has subsided and, so, are available to stimulate the appropriate immune response.

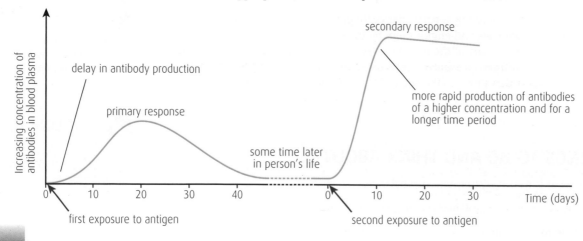

ALLERGY

Sometimes the body over reacts to a small trace of a harmless foreign substance and the person suffers an **allergic response**. The antigen involved is called an **allergen**. Typical allergens include animal hair, pollen, moulds and dust mites.

Allergy is a hypersensitive B-lymphocyte response to an antigen which is normally harmless. An allergic reaction involves the production of histamine, resulting in inflammation and damage to tissues.

Some common diseases caused by allergic reactions are hay fever, anaphylactic shock and allergic asthma.

Hay fever

Anaphylactic shock

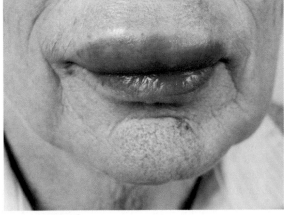

Allergic asthma

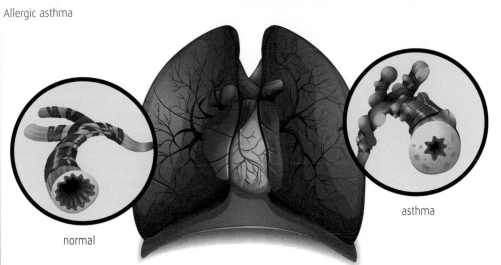

normal

asthma

ONLINE

Read about the causes and symptoms of hay fever at www.brightredbooks.net

ONLINE

Read about the causes and symptoms of anaphylactic shock at www.brightredbooks.net

ONLINE

Read about the causes and symptoms of allergic asthma at www.brightredbooks.net

VIDEO LINK

Watch the video about the allergic reaction at www.brightredbooks.net

ONLINE TEST

Take the test on specific cellular defence at www.brightredbooks.net

THINGS TO DO AND THINK ABOUT

1 a   Using the graph shown on p76 describe differences between the primary and secondary responses to infection by a particular antigen.
   b   Explain why the secondary response differs from the primary response.

2 Draw a diagram and label it to show:
   a   an antibody
   b   the antigen-binding sites
   c   the antigen–antibody complex.

3 a   Describe what is meant by an allergic response.
   b   What is the cause of this type of reaction?

SPECIFIC CELLULAR DEFENCE 3

CASE STUDIES

ABO Blood Types

There are four main blood groups in humans: A, B, AB and O. They are categorised by the presence or absence of two types of antigen (A and B) on the surface of a person's red blood cells.

Blood group	A	B	AB	O
Antigens on red blood cells	A antigens	B antigens	Both A and B antigens	Neither A nor B antigens
Antibodies in blood plasma	Anti-B antibodies	Anti-A antibodies	Neither anti-A nor anti-B antibodies	Both anti-A and anti-B antibodies

Antigens and antibodies linked to each blood group

As the table shows, antibodies are present in blood plasma and so certain combinations of blood types are not compatible, resulting in **agglutination** or clumping of the red blood cells.

If donor blood containing B antigens is given to a patient with either group A or group O blood, the antibodies present in the patient's plasma cause agglutination. Similarly, donor blood containing A antigens will agglutinate if given to a patient with either blood group B or group O. Blood group O is the universal donor since it has neither of the antigens on its red cells. Blood group AB is the universal recipient since it has neither of the antibodies in its plasma.

Rhesus Blood Types

Red blood cells sometimes have another antigen on their red cells known as antigen D (or RhD). If this is present, the blood group is Rhesus positive (Rh+) and if absent the blood group is Rhesus negative (Rh−). If Rh+ blood were to be given by transfusion to someone who was Rh−, the recipient would produce anti-D antibodies and subsequent transfusions could lead to agglutination. The greatest problem arises if a Rh− mother has a foetus which is Rh+. Normally, the placenta acts as an immune barrier but maternal and foetus blood can mix during birth or miscarriage. In this case, the mother's immune system will respond by producing anti-D antibodies. In subsequent pregnancies, if the mother is carrying a Rh+ foetus, it will be attacked by the anti-D antibodies. This is known as haemolytic disease of the newborn, which can now be prevented.

AUTOIMMUNE DISEASE

An autoimmune disease occurs when the regulation of the immune system fails and can no longer distinguish between healthy tissue and antigens. T lymphocytes start to respond to 'self' antigens and, so, healthy tissue is attacked, resulting in destruction of healthy body tissue, abnormal growth of an organ or changes in the function of an organ.

DON'T FORGET

Incompatibility is between the red cells of the **donor** and the plasma of the person receiving the blood (**recipient**).

VIDEO LINK

Watch the videos about blood groups, transfusion and agglutination at www.brightredbooks.net

DON'T FORGET

T lymphocytes allow the body to distinguish between 'self' and 'non-self' molecules, due to their specific surface proteins.

DON'T FORGET

Cytokines are chemicals which act as mediators between cells and are involved in the regulation of the immune response.

Rheumatoid Arthritis

Rheumatoid arthritis is an autoimmune disease that causes inflammation of joints, resulting in pain and swelling. It has been discovered that there are a large number of cytokines active in the joints of patients with rheumatoid arthritis and that these chemicals promote the disease. A loss of regulation, leading to an imbalance between pro- and anti-inflammatory cytokines, results in inflammation and the destruction of the joint.

Type 1 Diabetes

Type 1 diabetes occurs when the body does not produce enough insulin, a hormone which controls the glucose content of the blood. It is an autoimmune disease in which the immune system attacks and destroys the insulin-producing cells (beta cells) in the pancreas.

Joint damage as a result of rheumatoid arthritis

Multiple Sclerosis

Multiple sclerosis is an autoimmune disease which damages the myelin sheath, the protective fatty covering of nerve cells. The nerve damage is caused by inflammation whereby the body's own immune cells (T cells) attack antigens on the myelin sheath, resulting in its damage and so nerve impulses slow down or stop. This damage can occur in any area of the central nervous system including the brain, optic nerve and spinal cord.

THINGS TO DO AND THINK ABOUT

1 Look at the table on page 78 and explain why:
   a blood group O is called the universal donor
   b blood group AB is called the universal recipient.

2 Describe what causes an autoimmune disease.

ONLINE

Read more about rheumatoid arthritis at www.brightredbooks.net

VIDEO LINK

Watch the video about type 1 diabetes at www.brightredbooks.net

ONLINE

Read more about multiple sclerosis at www.brightredbooks.net

ONLINE TEST

Test yourself on specific cellular defence at www.brightredbooks.net

THE TRANSMISSION OF INFECTIOUS DISEASES

Globally, much current research in public health is focussed on the vital role played by the immune system in the human body, in fighting infectious diseases and maintaining good health.

DON'T FORGET

An infectious disease is caused by a pathogen and can be spread from one person to another.

ONLINE

Read about various ways in which infections can be transmitted at www.brightredbooks.net

ONLINE

Read about the transmission of the ebola virus at www.brightredbooks.net

INFECTIOUS DISEASES

Disease-causing pathogens include bacteria, viruses, fungi, protozoa and multicellular parasites.

Pathogen	Examples of diseases caused by this pathogen
Bacteria	food poisoning, cholera, typhoid, whooping cough, gonorrhoea
Viruses	influenza (flu), colds, measles, mumps, rubella, chicken pox, AIDS
Fungi	athlete's foot, ringworm
Protozoa	malaria, African sleeping sickness
Multicellular parasites	schistosomiasis

TRANSMISSION OF INFECTION

Infections can be transmitted from one person to another by several different means.

1 Direct Physical Contact

ONLINE

Read about current issues regarding diseases transmitted in water at www.brightredbooks.net

Direct contact is the most common way in which infections are transmitted; this occurs when an infected person makes physical contact with a healthy person and the pathogenic organism is passed on. For example, if you shake hands with someone who has the common cold and who has recently wiped their nose, they may pass the virus on to you; there may be cold virus particles on their hands. It is also possible for a pregnant woman to pass some infections directly to their unborn baby via the placenta or during childbirth. Some organisms can live on objects for a short time, so touching a door handle soon after an infected person may expose you to infection.

2 Water

If fresh water is contaminated with pathogenic microorganisms, infection may be transmitted during bathing, washing, drinking, or in the preparation or consumption of food. **Typhoid**, **dysentery** and **cholera** are transmitted in this way. Water can be contaminated by the faeces of animals or humans.

3 Food

ONLINE

Learn more about foodborne illnesses at www.brightredbooks.net

Pathogens can enter food chains; if humans eat food which has been contaminated, infection will be transmitted – particularly if food is eaten raw or is undercooked. Microbes can enter the gut and cause sickness and diarrhoea, such as in salmonella food poisoning. *E. Coli* bacteria are also transmitted via food. Microbes can spread from one food to another during food preparation, for example from the hands or on kitchen utensils. *Clostridium botulinum*, which causes **botulism** can be transmitted during the canning process.

ONLINE

Read about salmonella food poisoning at www. brightredbooks.net

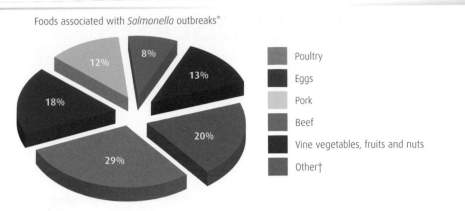

Foods associated with *Salmonella* outbreaks*

- Poultry
- Eggs
- Pork
- Beef
- Vine vegetables, fruits and nuts
- Other†

8% 13% 12% 18% 20% 29%

*These contaminated ingredients or single foods (belonging to one food category) were associated with one third of the Salmonella outbreaks.

†Other includes: Sprouts, leafy greens, roots, fish, grains-beans, shellfish, oil-sugar, and dairy.

4 Body Fluids

Body fluids, such as saliva, blood, semen and vaginal fluid, can contain microorganisms which may be transmitted from an infected person to others. Contact with infected blood can pass on **HIV**, **hepatitis B** or **hepatitis C**. Hepatitis B and HIV can be spread through sexual contact or via blood-contaminated syringe needles. Kissing can result in transmission of the cold or flu viruses.

5 Inhaled Air

Infectious diseases can also be spread, indirectly, through the air. Certain pathogens which are spread by coughs or sneezes can survive for a long time, suspended in the air in the residue from droplets which have evaporated or in dust particles. These pathogens are resistant to drying out. Organisms spread by airborne transmission are inhaled and so enter the respiratory tract and infect another person. The **common cold** can be spread by inhaling airborne viruses after someone has sneezed or coughed, and the **flu** virus is also spread by inhaling infected droplets in the air. **Measles**, **mumps**, **tuberculosis** and **chickenpox** can also be transmitted in this way.

6 Vector Organisms

A **vector** is a living thing which can pass a pathogenic microorganism from one living thing to another, such as from human to human, or animal to human. Blood-sucking insects, such as mosquitoes, fleas, and ticks, are common vectors and can ingest pathogens when they feed on an infected host. When the insect bites another host, the infection is passed on. **Malaria**, **yellow fever** and **Lyme disease** are spread via vectors as shown in the table.

Disease	Pathogen	Vector
Malaria	protozoan	mosquito
Lyme disease	bacteria	tick
Yellow fever	virus	mosquito

The malaria mosquito

THINGS TO DO AND THINK ABOUT

1 Research the following three infectious diseases and, for each, write a short paragraph explaining how the infection is transmitted from one person to another:
   a measles   b HIV   c cholera.

2 Complete the following table for three other diseases showing their mode of transmission. Each mode of transmission should be different.

Disease	Mode of transmission

DON'T FORGET

A parasite lives in or on the body of another organism which is known as the **host**, and gains energy or nutrients from it.

ONLINE TEST

Head to www. brightredbooks.net to test your knowledge of the transmission of infectious diseases.

CONTROL OF INFECTIOUS DISEASES

There are several strategies which can prevent the spread of an infectious disease.

STRATEGIES TO CONTROL INFECTION

1 Quarantine

An individual who is known to have a particular infection may be isolated from healthy people, in order to prevent the disease spreading. **Quarantine** is when a person who is known to have been **exposed** to a particular infection is kept in compulsory isolation for a period of time (the known incubation period for that particular type of infection). The aim of a period of quarantine is to prevent spread of infection. It separates those who are known to have been exposed (and who may or may not become ill) from those who are known to be healthy, in the hope of stopping further spread of disease.

2 Antisepsis

Prevention of the spread of infection is minimised during surgery by using sterile procedures: wearing a sterile gown, gloves and a mask, and all equipment is sterilised using heat treatment (boiling in water or autoclaving using steam). An antiseptic is a chemical which kills microorganisms and can be applied to the skin to reduce the chance of infection. Common antiseptics include alcohol, hydrogen peroxide and iodine, and there are many commercial liquids, creams, sprays and wipes available.

3 Individual Responsibility

Every individual can help prevent the spread of infection by being responsible for good hygiene, care in sexual health and through appropriate storage and handling of food.

Wash Your Hands!

Individual responsibility	Explanation/examples
Good hygiene	Washing hands, showering and the use of antibacterial hand gels can prevent transmission of pathogens from one host to another.
Care in sexual health	Using condoms during sex can prevent the spread of sexually transmitted infections.
Appropriate storage and handling of food	Keeping work surfaces, hands and utensils clean during preparation of food. Cooking food thoroughly. Only reheating food once. Storing food at the correct temperature. Storing raw and cooked food separately.

4 Community Responsibility

Larger communities, such as countries and local authorities, or smaller communities, such as towns or villages, can help prevent the spread of infection by ensuring a safe water supply, and by having appropriate waste disposal systems. Food webs must be kept free from harmful microorganisms during cultivation, harvesting and manufacturing processes, so that the food we eat is safe.

Water treatment plant

Sewage treatment plant

Community responsibility	Explanation/examples
Quality water supply	Water is **filtered** to remove any solid particles and treated with **chlorine** to kill pathogens.
Safe food webs	Milk is **pasteurised** to kill many microbes. Businesses which make or prepare food are **inspected** for hygiene, must carry out risk analysis of food and must ensure **traceability** (be able to identify their suppliers of food and the businesses to which they supply products).
Appropriate waste disposal systems	Waste water (sewage) is kept separate from the fresh water supply and is treated to kill harmful microbes. Dry waste is collected regularly and recycled, incinerated or buried.

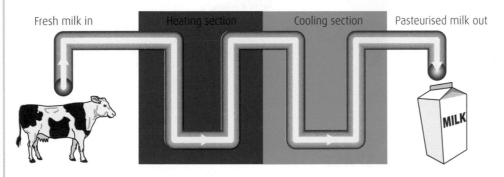
Fresh milk in — Heating section — Cooling section — Pasteurised milk out

MILK

5 Vector Control

Controlling the vectors of disease can prevent the spread of infection. Malaria is a serious tropical disease caused by a protozoan parasite, the plasmodium, which is spread by the mosquito vector. When an infected mosquito bites a human, it passes the parasites into the bloodstream. Removing the mosquito vector is a key strategy in the control and eradication of malaria. This can be achieved by:

1   draining stagnant water to remove mosquito breeding grounds

2   introducing sterile male mosquitoes to reduce breeding rates

3   using chemicals such as insecticides and larvicides.

THINGS TO DO AND THINK ABOUT

1   Describe the meanings of: (i) quarantine and (ii) antisepsis.
2   Explain two ways in which you, as an individual, can prevent spread of infection by showing responsibility.
3   Explain two ways in which communities can act responsibly to prevent the spread of infection.
4   Explain what is meant by vector control.

EPIDEMIOLOGICAL STUDIES OF INFECTIOUS DISEASE

Epidemiologists study disease outbreaks and patterns of infection in order to determine the factors which affect the spread of disease. They may focus on the **area** where the initial outbreak occurred, the **pattern and speed** of spread, and the **geographical distribution** of the infection. This data allows appropriate control measures to be considered and put into action.

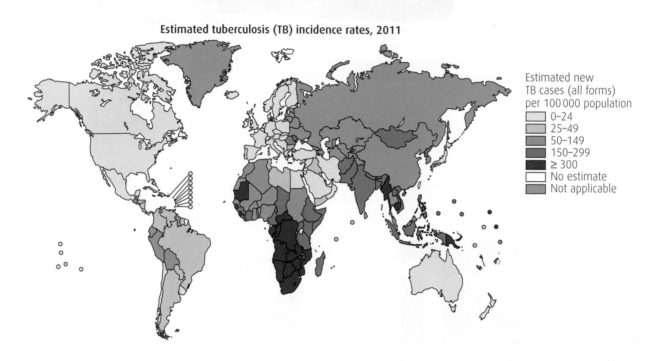

Estimated tuberculosis (TB) incidence rates, 2011

Estimated new
TB cases (all forms)
per 100 000 population
- 0–24
- 25–49
- 50–149
- 150–299
- ≥ 300
- No estimate
- Not applicable

SPREAD OF DISEASE

The spread of a disease in a particular geographical location can be described as either sporadic, endemic, epidemic or pandemic.

- **Sporadic** refers to a disease that occurs infrequently and irregularly (occurring only occasionally in isolated areas and where outbreaks are not connected).

- **Endemic** describes the regular, observed, baseline level of a disease that is constantly present in a geographical area (not necessarily zero). It is the usual prevalence of a disease; for example malaria is endemic to parts of Africa.

- **Epidemic** describes an increase in the number of cases of a disease above what is normally expected in the population of that area – an unusually high number of cases. The disease attacks many people in a given area in a short period of time and may spread through several communities, for example, influenza.

- **Pandemic** is a series of epidemics which spread over several countries, continents or the world and which affect a large number of people. It may be described as a global epidemic.

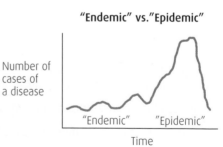

"Endemic" vs."Epidemic"

Number of cases of a disease

"Endemic"   "Epidemic"

Time

CONTROL MEASURES

The main purpose of an investigation of an outbreak of an infectious disease by an epidemiologist is to determine its **cause**, halt its **spread** and put in place measures to **minimise the risk of further outbreaks**. Effective control measures are vital for stopping the outbreak and preventing reoccurrence. Examples of the main control measures used, sometimes in combination, are shown in the table.

Control measure	Example(s)
Prevention of transmission	Quarantine, good hygiene, antisepsis, care in sexual health, care in storage and handling of food, safe water supply, food webs and waste disposal, control of vectors (see page 83)
Drug therapy	Antibiotics for treatment of bacterial infections, some anti-virals are available, anti-malarial medication to prevent or cure malaria
Immunisation	People can be made resistant to a particular disease by vaccination; vaccines stimulate the immune system to protect against a subsequent infection

ONLINE

Learn more about immunisations and viruses at www.brightredbooks.net

DON'T FORGET

Immunity is the ability of the body to resist infection by a pathogen or to destroy it if it is successful in invading the body

ACTIVE IMMUNISATION AND VACCINATION

Active Immunity

Immunisation is a process whereby a person develops **immunity** to a particular pathogenic microorganism. **Active immunity** refers to a type of immunity in which antibodies are produced by the person's own body, in response to exposure to an antigen. Active immunity can be stimulated by **vaccination**. This is when a **vaccine** (an antigenic material from an infectious pathogen) is administered, which

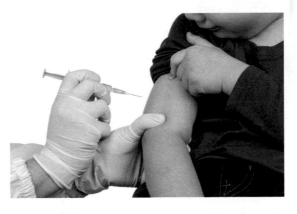

stimulates a person's immune system to produce antibodies against the particular antigen and to develop immunological memory (see page 76). The antigens from the infectious pathogen are usually mixed with an **adjuvant** in order to enhance the immune response. Types of antigen which are used in vaccination are shown in the table.

Type of antigen used in a vaccine	Examples of diseases for which this type of antigen is used as a vaccine
Inactivated pathogen toxins	Tetanus, diphtheria
Dead pathogens	Polio, hepatitis A
Parts of pathogens	HPV, hepatitis B
Weakened pathogens	Measles, mumps, rubella

DON'T FORGET

Antibodies are produced by B lymphocytes. Specific antibodies recognise and combine with specific antigens.

DON'T FORGET

An adjuvant is a pharmacological agent which is added to a drug to increase or aid its effect.

THINGS TO DO AND THINK ABOUT

1 Explain the meaning of the following terms:
   a epidemiologist
   b endemic, epidemic and pandemic disease
   c active immunity
   d vaccination.

2 Describe two different types of antigen that are used as vaccines and name the diseases they protect against.

ONLINE TEST

Take the test on epidemiological studies of infectious disease at www.brightredbooks.net

VACCINE CLINICAL TRIALS AND HERD IMMUNITY

VACCINE CLINICAL TRIALS

All new pharmaceutical products (including vaccines) must undergo **clinical trials** before being licensed for use. After a vaccine has been developed and tested in a lab, clinical trials (in humans) are needed to establish **safety**, determine **effectiveness** as a treatment and to discover any **side effects**.

DESIGN OF CLINICAL TRIALS

As in all scientific investigations, a vaccine clinical trial must use procedures that ensure a fair and valid trial, including:

1 Randomisation

In a **randomised** clinical trial, there is an experimental group of people who are given the new treatment and a control group who are given no treatment or a placebo. Participants are assigned randomly (by chance) to each group.

2 Placebo

To assess a vaccine, it must be compared to an alternative treatment, for example a previously licensed vaccine or an inactive substance. A **placebo** is a 'dummy' treatment with no active ingredient.

3 Double-blind

In a **double-blind** study, to **eliminate bias**, neither the investigator nor the participant is aware of the nature of the treatment the participant is receiving. The expectations of the researcher and the participant do not affect the outcome. One group of subjects would receive the vaccine while the second group would receive the placebo control. This would ensure a valid comparison and help determine the effectiveness of the vaccine.

4 Sample size

The **sample size** (the number of participants assigned to the control and the experimental groups) affects the reliability of the results of a clinical trial. The larger the group size, the smaller the magnitude of experimental error and the greater the statistical significance of the results.

5 Analysis of results

Once a study is completed, the results from the experimental group and the control group are compared to determine if there is a **statistically significant difference** between treatment and control.

ONLINE

Find out about the design of clinical trials and their use in the development of vaccines at www.brightredbooks.net

HERD IMMUNITY

Herd immunity occurs when a **significant proportion** of a population has been vaccinated against a particular infection, providing a degree of protection for those who are not immune. If a high proportion of the population is immunised, this lowers the probability that non-immune individuals will come into contact with infected individuals and the disease is less likely to spread.

contd

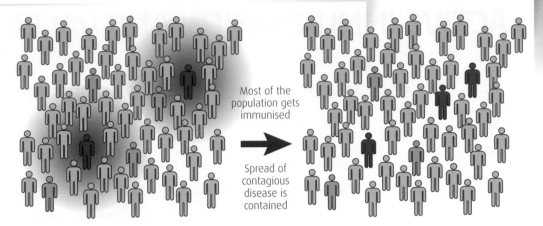

- **immunised and heathly**
- **not immunised but still healthy**
- **not immunised, sick and contagious**

Most of the population gets immunised

Spread of contagious disease is contained

The Importance of Herd Immunity

Herd immunity is important in reducing the spread of diseases in a community, but can also be crucial for protecting vulnerable groups of people who cannot be vaccinated, for example:

- people with a weakened immune system, such as those undergoing chemotherapy
- new-born babies and children who are too young to be vaccinated
- elderly people
- HIV-infected people
- people who are seriously ill.

Herd Immunity Threshold

The **herd immunity threshold** is the proportion of individuals in a population who must be immune to prevent a particular disease spreading. Instead the disease will decrease to a low baseline level. This threshold depends on several factors.

Herd immunity thresholds for selected vaccine-preventable diseases

Disease	Herd threshold immunity
Diphtheria	85%
Measles	83–94%
Mumps	75–86%
Pertussis	92–94%
Polio	80–86%
Rubella	83–85%
Smallpox	80–85%

Factor affecting herd immunity threshold	Explanation
The particular disease	The proportion of the population which must be immunised in order to achieve herd immunity is different for different diseases. More virulent diseases require a greater percentage of vaccination in the population to give the desired herd immunity.
The efficacy of the vaccine used in immunisation	The effectiveness of the vaccine affects the threshold. This can be determined using a clinical trial.
The contact parameters for the population	The number of contacts made with other individuals in a population in a certain period of time affects the threshold. This will depend on the population density.

If immunisation rates fall, herd immunity can break down, resulting in an increase in new cases of infection. Herd immunity eradicated smallpox and explains why diseases such as polio and diphtheria are rare in developed nations where there are well-established vaccination programmes.

MASS VACCINATION

Mass vaccination is the immunisation of a large number of people in a particular location at one time or within a short time interval.

 THINGS TO DO AND THINK ABOUT

Investigate the effect of mass vaccination on the following three diseases:

1 TB   2 Polio   3 Smallpox

 VIDEO LINK

Read about herd immunity and watch the videos at www.brightredbooks.net

 DON'T FORGET

Vaccination is the administration of a vaccine in order to stimulate the immune system to develop immunity to a pathogen.

ONLINE

Read about the eradication of smallpox through herd immunity at www. brightredbooks.net

 DON'T FORGET

Take the test on this topic at www.brightredbooks.net

ACTIVE IMMUNISATION AND VACCINATION

ONLINE

Read about the global vaccine action plan and the childhood immunisation programme in Scotland at www.brightredbooks.net

DON'T FORGET

The aim of any vaccination programme is to reduce or eliminate a particular disease.

DON'T FORGET

The herd immunity threshold is the proportion of individuals who must be immune so that a particular disease will no longer spread through the population, but decrease to low incidence.

ONLINE

Read about the effects of the MMR vaccination scare on the increase in measles at www.brightredbooks.net

DON'T FORGET

Immune surveillance refers to white blood cells which monitor for tissue damage and invasion by pathogens.

DON'T FORGET

Bioinformatics is the use of computer technology to identify DNA sequences.

PUBLIC HEALTH IMMUNISATION PROGRAMMES

In most countries, the Policy for Public Health aims to establish herd immunity to a number of diseases. The population of Scotland is protected through immunisation against a number of infectious diseases. As shown in the table, each of these diseases has a different herd immunity threshold.

Diseases for which vaccination is routine in Scotland	Herd immunity threshold, % (where available)
diphtheria	85
whooping cough	92–94
polio	80–86
Measles, mumps and rubella (MMR)	Measles 83–94 Mumps 75–86 Rubella 83–85
tetanus	N/A *
Haemophilus influenzae type b (Hib)	Data unavailable
Meningococcal C disease (Men C)	Data unavailable
Pneumococcal infection	Data unavailable
Human Papilloma Virus (HPV)	Data unavailable

*Herd immunity only applies to diseases that are contagious. It does not apply to diseases such as tetanus where the vaccine protects **only** the vaccinated person.

Herd immunity is difficult to establish in some populations. In **developing countries** widespread vaccination may not be possible due to malnutrition and poverty. Poverty prevents access to health services since vaccination may not be affordable for all.

In **developed countries**, vaccinations may be rejected by a percentage of the population. Some parents choose not to have their children immunised due to fears over the safety or possible side effects of vaccines.

THE EVASION OF THE SPECIFIC IMMUNE RESPONSE BY PATHOGENS

The human immune system has developed specific mechanisms to defend the body against pathogens. However, pathogens have also developed various strategies, through the process of evolution, to evade the immune system. This has consequences for vaccination strategies.

Antigenic Variation

Some pathogens evade immune surveillance and so avoid the effect of immunological memory by changing their surface proteins (antigens). Different antigens form as a result of mutations in the pathogen's DNA. This is **antigenic variation**. Antibodies in the body no longer recognise the pathogen since memory cells produced from an initial infection fail to recognise the new antigens. The current vaccine will not protect against the new form of the disease. Antigenic variation also makes it difficult to develop new vaccines for certain infections.

contd

The Impact of Antigenic Variation in Disease

Antigenic variation occurs in diseases such as **malaria** and **trypanosomiasis**, which is one of the reasons why they are still common today in many parts of the world.

Influenza

Antigenic variation also occurs in the influenza virus. Therefore, the memory cells produced due to an infection by one strain of the flu virus will not recognise the antigens of a new strain. This explains why flu is still a major public health problem and why those at risk from complications from influenza must be vaccinated annually.

Bioinformatics software can be used to study the differences in DNA or protein sequences between different strains of influenza viruses.

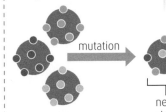

Distribution of trypanosomiasis in Africa

Incidence in local population, per year
No cases
<100 cases
100–1000 cases
>1000 cases

Risk for travellers
(Cumulative reported cases 1983–2008)
+ <10 infections in travellers per country
++ ≥10 infections in travellers per country

Uganda: overlap *T b gambiense* and *T b rhodesiense* possible

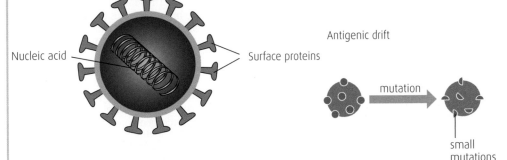

The diagram below shows the structure of one strain of the influenza visus

Nucleic acid

Surface proteins

Antigenic drift

mutation

small mutations

Antigenic shift

mutation

new strain

Antigenic variation in the influenza virus

ONLINE

Learn more about antigenic variation at www. brightredbooks.net

DIRECT ATTACK ON THE IMMUNE SYSTEM

A direct attack on the immune system, which removes a function or causes part of it to fail, will increase the likelihood of infection.

HIV

HIV (human immunodeficiency virus) directly attacks the immune system, destroying lymphocytes, particularly the T cells, and so increases susceptibility to infection. HIV is the major cause of AIDS in adult humans, and occurs when the body no longer has the ability to fight infections. This is life threatening.

Tuberculosis

Tuberculosis (TB) is a highly infectious bacterial disease which affects the lungs. The bacterium can persist in some people, since it can survive *within* phagocytes and so avoids detection by the immune system.

 VIDEO LINK

Find out about HIV infection at www.brightredbooks.net

 THINGS TO DO AND THINK ABOUT

1 Research the public health measures and drug therapies used to control HIV infection.
2 Explain why herd immunity may not be successfully established in (i) a developing country and (ii) a developed country.
3 Explain what is meant by antigenic variation.
4 A different vaccine is needed for each strain of the influenza virus. Why are different vaccines required?

 ONLINE TEST

Head to www. brightredbooks.net and test your knowledge of this topic

APPENDICES

ANSWERS

UNIT 1: HUMAN CELLS

Division and differentiation in human cells, pp.6–7

1 Vital and specialised genes

2 Pluripotent: able to form any type of specialised cell; multipotent: ability to form specialised cell is limited those found in the tissue from which they arise.

3 To provide a pool of stem cells for future needs.

Stem cells, pp.8–9

1 Testes and ovaries

2 a More germline cells

   b Gametes

3 Rapid, out-of-control cell division forms a tumour; a good blood supply forms to provide nutrients; the cells no longer stick to each other so can be carried away by the blood circulation (metastasis).

DNA, pp.10–11

1 16:9

2 a 1 deoxyribose, 2 phosphate group, 3 cytosine, 4 adenine

   b
   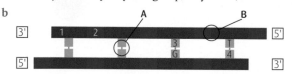

   c A: weak hydrogen; B strong chemical

3 Complete this diagram to show the direction of DNA replication. Show the 3' and 5' ends and draw the leading strand with a continuous line and the lagging strand with a broken line.

Gene expression, pp.12–13

1

	DNA	mRNA
Type of sugar	Deoxyribose	Ribose
Bases	Adenine, cytosine, guanine and thymine	Adenine, cytosine guanine and uracil
Number of strands	Two	One
Location	Only in nucleus	Moves from nucleus to cytoplasm

2 mRNA carries transcribed code from DNA to ribosome; tRNA anticodon matches with mRNA codon, bringing a specific amino acid into the correct position in a sequence; rRNA binds with proteins to form ribosomes which are the site of protein translation.

3
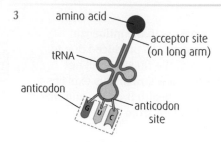

Protein synthesis, pp.14–15

1 UCCGAUUGACGUUAGCUUUAC

2 Endonuclease, ligase, primer, free mRNA nucleotides, ATP, DNA template strand

3 1 DNA unwinds as RNA Polymerase moves along a section that codes for a protein. 2 The DNA molecule 'unzips' when hydrogen bonds are broken. 3 Bases on the DNA strands are exposed. 4 mRNA nucleotides move in and form complementary base pairs with one of the DNA strands (the coding strand). Weak hydrogen bonds form. Cytosine always pairs with guanine; adenine on DNA pairs with uracil on mRNA, and thymine on DNA pairs with adenine on mRNA. 5 Strong chemical bonds form between the phosphate of one nucleotide and the ribose of the next nucleotide, building the mRNA strand. 6 The weak hydrogen bonds that were holding the DNA and mRNA strands together break, allowing the mRNA primary transcript to leave the nucleus and enter the cytoplasm. 7 Hydrogen bonds reform between the two DNA strands, and the DNA molecule rewinds to form a double helix.

4 Non-coding regions = introns, coding regions = exons

5 Primary transcript is longer as it contains introns.

One gene, many proteins, pp.16–17

1 a Peptide

   b Hydrogen

   c Primarily hydrogen with contributions from hydrophobic and ionic bonds.

2 Introns are spliced out but the same nucleotide sequence may be cut differently to produce different proteins.

3 Other non-protein groups such as phosphate or carbohydrate groups may be added, altering function. The polypeptide chain may be shorter if sections are removed or bigger if other polypeptide chains are added.

Mutations, pp.18–19

1 a translocation   b duplication   c deletion

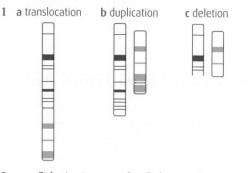

2 a Substitution   b Deletion   c Insertion

Polymerase chain reaction (PCR), pp.22–23

1   Heating to 92°C separates the DNA strands, at 55°C primers can bind to the strands.

2   It would be denatured in the PCR process and amplification would not occur.

Metabolism, pp.24–25

1   a   Regulator codes for the synthesis of a repressor protein.

   b   Operator switches on the structural gene.

   c   Structural codes for the synthesis of the enzyme (β-galactosidase)

2   Saves resources and energy

Enzyme action, pp.26–27

1   The active site of the enzyme is open and the substrate moves towards it. The active site closes bringing the substrate closer in and causing the reaction to occur. The resultant product triggers the opening of the active site and it is released.

2   Competitive: the inhibitor binds to the active site, preventing substrate from getting near. Non-competitive: the inhibitor binds to some other part of the enzyme, but changes the shape of the active site, thus preventing it binding with the substrate.

3   The shape of the active site is altered so no amount of substrate can react with it.

Cellular respiration 1, pp.28–29

1   It catalyses an irreversible reaction which commits the resultant molecule to glycolysis.

2   Removes **hydrogen** and **electrons** from respiratory intermediates and passes them to the hydrogen carriers NAD (to form $NADH_2$) and FAD (to form $FADH_2$).

Cellular respiration 2, pp.30–31

When ATP or citrate increases to an excessive level they can each inhibit the action of photofructokinase. This slows down glycosis. Eventually the levels of ATP and/or citrate drops and this lowers the inhibitory effect; glycolysis, and consequently the citric acid cycle, speeds up to meet the energy demand.

Exercise, pp.32–33

1   They have a richer blood supply

2   Fast twitch 35: slow twitch 65

3

	Slow twitch	Fast twitch
Number of mitochondria	Many	Few
Blood supply	Rich	Poor
Source of energy	Fat	Glycogen, creatine phosphate
Type of activity	Endurance: cycling, long distance running	Short bursts: javelin, shot putt

UNIT 2: PHYSIOLOGY AND HEALTH

Male reproductive system, pp.34–35

(i)   Testosterone level is unaffected because, like other hormones, it is transported in the blood.

(ii) Sperm production continues, but the sperm cannot pass into the urethra.

Cardiovascular system, pp.44–45

1   (a) Elastic fibres allow the artery to stretch as blood pulses through. (b) Smooth muscle in arteriole walls contracts to narrow the vessel.

2   Veins have valves to prevent backflow of blood.

3   Blood plasma contains proteins which are not present in tissue fluid.

The heart, pp.46–47

B

Pathology of cardiovascular disease (CVD) 2, pp.52–53

B

Blood glucose level and diabetes, pp.54–55

1   Pancreas

2   X: insulin

   Y: glycogen

3   Liver

4   Glucose would be used up in respiration.

Obesity, pp.56–57

The percentage of obese boys between 6 and 19 years of age increases from 1971–2000.

The percentage of obese girls between 6 and 19 years of age increases from 1971–2000.

UNIT 3: NEUROBIOLOGY AND COMMUNICATION

Perception, pp.60–61

1   B

2   D

Nervous system, pp.64–65

Glial cells physically support the neurons and provide them with nutrition. They also remove debris by phagocytosis.

Behaviour, pp.68–69

1   a   To prevent cheating by using more than one finger.

   b   To increase validity by controlling variables.

   c   To prevent cheating by looking at the maze.

   d   Enough trials must be included to allow learning to take place.

   e   Increases reliability.

2   Social facilitation

3   Generalisation

4   D

Communication and social behaviour, pp.70–71

1   A long period of dependency allows the development of social and communication skills.

2   (i)   The infant cannot communicate verbally, so relies on non-verbal communication.

   (ii) Clapping, smiling.

ANSWERS (CONT)

UNIT 4: IMMUNOLOGY AND PUBLIC HEALTH

Immunology, pp.72–73

1   Epithelial cells form an outer layer that provides a physical barrier against the entry of pathogens. They also contain specialised cells which produce secretions that trap pathogens, e.g. wax, oil or mucus. Other glands produce antimicrobial chemicals.

2   a   Mast cells release histamine in damaged tissues.

   b   Histamine increases the permeability of capillaries and causes blood vessels to dilate, thus increasing blood flow to the site of infection.

   c   Phagocytes engulf, digest and destroy pathogens by phagocytosis.

   d   Platelets and other blood clotting factors seal the wound and so trap pathogens.

3   Apoptosis is programmed cell death. Natural killer cells release toxic proteins which perforate the membrane of e.g. cancer cells or cells infected with viruses. This allows signal molecules to enter the cells which stimulate the cells to produce self-destructive enzymes.

Specific cellular defence 1, pp.74–75

1   Immune surveillance is a theory which states that a wide range of white blood cells are constantly circulating the body to monitor for tissue damage or infection by pathogenic microorganisms. Immune surveillance may also detect cancer cells.

2   The body has many different lymphocytes, each with different membrane receptors which are specific for one particular antigen. The clonal selection theory suggests that an antigen selects and binds to its receptor on the specific lymphocyte. The lymphocyte then undergoes repeated division, resulting in a clone of identical lymphocytes.

3

T lymphocyte type	Function
Cytotoxic T cells	Destroy foreign cells by apoptosis
Helper T cells	Secrete cytokines which activate B lymphocytes and phagocytes
Memory T cells	Involved in immunological memory

Specific cellular defence 2, pp.76–77

1   a   The secondary response (i) produces antibodies more quickly, (ii) produces a higher concentration of antibodies and (iii) is longer lasting than the primary response (the antibodies remain in the blood for longer).

   b   These differences occur because of immunological memory. Some T and B lymphocytes which are produced in response to exposure to antigens survive as memory cells. Therefore, the immune system can respond more quickly and more effectively if exposed to the same antigen in the future.

2
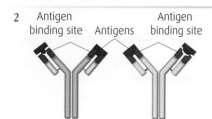
Antigen binding site   Antigens   Antigen binding site

3   a   An allergic response occurs when the body overreacts to a small trace of a harmless foreign substance.

   b   Histamine is produced in response to the allergen, resulting in inflammation and damage to tissues.

Specific cellular defence 3, pp.78–79

1   a   Blood group O is called the universal donor because there are no antigens on the red blood cells and so this blood type can be given to all other blood groups.

   b   Blood group AB is called the universal recipient because there are neither anti-A nor anti-b antibodies in the plasma and so these patients can safely receive blood from all other groups.

2   An autoimmune disease occurs when the immune system fails to distinguish between healthy cells and antigens. T Lymphocytes respond to self-antigens and, so, healthy tissues are attacked or function abnormally.

The transmission of infectious diseases, pp.80–81

1   a   The measles virus is airborne and is released from an infected person during coughing, sneezing or via exhaled air and is transmitted to another person when infected droplets are inhaled.

   b   The human immunodeficiency virus is transmitted from one person to another in body fluids. It can be passed on through sharing of needles by infected drug addicts and during sexual intercourse via semen.

   c   Cholera bacteria are transmitted if contaminated water is used for drinking water or in the preparation of food.

2   Any appropriate example of a disease and the correct mode of transmission, for example:

Disease	Mode of transmission
Diphtheria/rubella/whooping cough	Airborne droplets
Cold and flu	Direct contact or inhaled droplets
Botulism/salmonella	Contaminated food
Hepatitis	Body fluids
Malaria	Insect vector

Control of infectious diseases, pp.82–83

1   (i)   Quarantine is when someone who is known to have been in contact with a particular infection is kept in compulsory isolation for a period of time (the incubation period of that infection) in order to prevent the spread of infection.

(ii) Antisepsis refers to strategies to minimise the spread of infection, e.g. wearing sterile gown, mask and gloves during surgery, using heat to sterilise equipment and using antiseptics to kill microorganisms on hands and work surfaces.

2 (Any two) ways in which individuals can prevent spread of infection by showing responsibility are:

- good hygiene, e.g. washing hands effectively, using antibacterial hand gels
- care in sexual health, e.g. using condoms to prevent spread of sexually transmitted infections
- appropriate storage and handling of food, e.g. keeping utensils, hands and work surfaces clean, cooking food thoroughly, storing raw and cooked food separately.

3 (Any two) ways in which communities can prevent spread of infection by acting responsibly are:

- ensuring a safe water supply so that it is free from pathogens
- keeping food webs safe, e.g. via pasteurisation of milk, inspection of businesses which prepare and make food and ensuring traceability of foods
- having appropriate waste disposal systems, e.g. the treatment of sewage so that it does not contaminate fresh water supplies and the regular collection and safe disposal of dry waste.

4 Vector control is the control of living things which can pass on pathogens to humans. An example is the removal of mosquitoes using several strategies in order to halt the spread of malaria.

Epidemiological studies of infectious disease, pp.84-85

1 a An epidemiologist studies outbreaks and patterns of infectious diseases so that they can determine the factors which affect the spread of infection.

b Endemic refers to the regular baseline level of a disease that is constantly present in an area.

Epidemic refers to the increase in the number of cases of a disease above the expected baseline level, resulting in an unusually high number of cases.

Pandemic refers to a series of epidemics spread over several countries, continents or the world, affecting a large number of people.

c Active immunity is a type of immunity whereby the antibodies are produced by the person's own body in response to exposure to an antigen.

d Vaccination is when a person is given a vaccine which is an antigen from a pathogen. The vaccine stimulates the immune system to produce antibodies against the antigen, resulting in immunological memory.

2 Any two from the following:

Type of antigen used as vaccine	Example disease
Inactivated pathogen toxin	Tetanus, diphtheria
Dead pathogens	Polio, hepatitis
Parts of pathogens	HPV, hepatitis B
Weakened pathogens	Measles, mumps, rubella

Vaccine clinical trials and herd immunity, pp.86-87

The effects of mass vaccination on:

1 Tuberculosis (TB). Mass vaccination introduced in UK in 1950s has resulted in herd immunity and so now not everyone is vaccinated.

2 Polio. Mass vaccination has also resulted in herd immunity and polio is more or less absent from developed countries. Babies are routinely immunised at the age of 2 months.

3 Smallpox has been completely eradicated worldwide.

Active immunisation and vaccination, pp.88-89

1 Control of HIV infection

Public health measures	Drug therapies
Education about how virus is spread, promotion of safe sex, i.e. use of condoms, supplying drug addicts with clean needles	There are several different antiretroviral drugs which stop HIV from replicating in the body at different stages of its life cycle. Other drugs are used to treat any associated infections.

2 (i) Herd immunity may not be established in a developing country because widespread vaccination may not be possible due to poverty and/or malnutrition.

(ii) Herd immunity may not be established in a developed country because vaccinations may be rejected by some members of the population, due to fears over safety of the vaccine.

3 Antigenic variation occurs when certain pathogens evade the immune response by changing their surface antigens. Therefore, the antibodies present will not be able to recognise the altered antigens and so will not attack the pathogen.

4 Different vaccines are needed for each strain of the influenza virus because antigenic variation occurs in the influenza virus. The memory cells produced due to an infection by one strain of the virus will not recognise the antigens of the new strain.

INDEX